History of Rocketry and Astronautics

AAS History Series, Volume 11

International Academy of Astronautics History Symposia

Front Cover Illustration:

Offical NASA Marshall Space Flight Center (MSFC) protrait photo of aerospace pioneer, Dr. Wernher von Braun (23 March 1912 - 16 June 1977). He was director of MSFC 1960-1970. MSFC photo 2-2747-2, taken by Fabian Bachrach (released 21 September 1962).

History of Rocketry and Astronautics

Proceedings of the Fifteenth and Sixteenth
History Symposia of the International Academy of Astronautics

Rome, Italy, 1981
Paris, France, 1982

Roger D. Launius, Volume Editor

R. Cargill Hall, Series Editor

AAS History Series, Volume 11

A Supplement to Advances in the Astronautical Sciences

IAA History Symposia, Volume 6

AAS Publications Office
P.O. Box 28130
San Diego, California 92198

Affiliated with the American Association for the Advancement of Science
Member of the International Astronautical Federation

First Printing 1994

ISSN 0730-3564

ISBN 0-87703-382-X (Hard Cover)
ISBN 0-87703-383-8 (Soft Cover)

Published for the American Astronautical Society
by Univelt, Incorporated, P.O. Box 28130, San Diego, California 92198

Printed and Bound in the U.S.A.

American Astronautical Society History Committee Members

1994

Dr. Roger D. Launius, Chair
National Aeronautics and Space
Administration History Office

Dr. Donald R. Baucom
BMDO/Historian, The Pentagon

Mr. Michael L. Ciancone
NASA Lewis Research Center

Mr. Stephen E. Doyle
Aerojet Propulsion Division

Dr. Donald C. Elder
University of Redlands

Dr. Thomas Fuller
Peterson Air Force Base

Mr. R. Cargill Hall
Center for Air Force History
Bolling Air Force Base

Dr. Michael Q. Hooks
Jet Propulsion Laboratory

Mr. Robert H. Jacobs
Univelt, Incorporated

Ms. Cathleen S. Lewis
National Air and Space Museum

Dr. Rick W. Sturdevant
Peterson Air Force Base

Dr. Shirley Thomas
Author

Dr. J. Edson Way
International Space Hall of Fame

AAS History Series

Volume 1 *Two Hundred Years of Flight in America: A Bicentennial Survey*, Edited by Eugene M. Emme, 1977, 326p, Third Printing 1981, Hard Cover $35; Soft Cover $25; special price for classroom text or bulk purchase.

Volume 2 *Twenty-Five Years of the American Astronautical Society: Historical Reflections and Projections*, 1954-1979, Edited by Eugene M. Emme, 1980, 248p, Hard Cover $25; Soft Cover $15.

Volume 3 *Between Sputnik and the Shuttle: New Perspectives on American Astronautics*, 1957-1980, Edited by Frederick C. Durant, III, 1981, 350p, Hard Cover $40; Soft Cover $30.

Volume 4 *The Endless Space Frontier: A History of the House Committee on Science and Astronautics*, By Ken Hechler, Abridged and edited by Albert E. Eastman, 1982, 460p, Hard Cover $45; Soft Cover $35.

Volume 5 *Science Fiction and Space Futures: Past and Present*, Edited by Eugene M. Emme, 1982, 278p, Hard Cover $35; Soft Cover $25.

Volume 6 *First Steps Toward Space*, Edited by Frederick C. Durant, III and George S. James, 1986, 318p, Hard Cover $45; Soft Cover $35.

Volume 7 *History of Rocketry and Astronautics*, Edited by R. Cargill Hall, 1986, Part I, 250p, Part II, 502p, sold as a set, Hard Cover $100; Soft Cover $80.

Volume 8 *History of Rocketry and Astronautics*, Edited by Kristan R. Lattu, 1989, 368p, Hard Cover $50; Soft Cover $35.

Volume 9 *History of Rocketry and Astronautics*, Edited by Frederick I. Ordway, III, 1989, 330p, Hard Cover $50; Soft Cover $35.

Volume 10 *History of Rocketry and Astronautics*, Edited by Å. Ingemar Skoog, 1990, 330p, Hard Cover $60; Soft Cover $40

Volume 11 *History of Rocketry and Astronautics*, Edited by Roger D. Launius, 1994, 236p, Hard Cover $60; Soft Cover $40

Volume 12 *History of Rocketry and Astronautics*, Edited by John L. Sloop, 1991, 252p, Hard Cover $60; Soft Cover $40

Volume 13 *History of Liquid Rocket Engine Development in the United States 1955-1980*, Edited by Stephen E. Doyle, 1992, 176p, Hard Cover $50; Soft Cover $35

Volume 14 *History of Rocketry and Astronautics*, Edited by Tom D. Crouch, Alex M. Spencer, 1993, 222p, Hard Cover $50; Soft Cover $35

Volume 15 *History of Rocketry and Astronautics*, Edited by Lloyd H. Cornett, Jr., 1993, 452p, Hard Cover $60; Soft Cover $40

These volumes are available to individual members of space and astronautical societies and students at half the list price.

Order from Univelt, Incorporated, P.O. Box 28130, San Diego, California 92198

Foreword

As subjects of world-wide significance, it seems to me that rocketry and astronautics have attracted far too many uncritical publicists of the "gee whiz" stripe and, with some notable exceptions, far too few serious scholars. Perhaps this is owed to the subject matter and its 20th century connection to national pride and international prestige. Whatever the case, the literature is rife with misperceptions and myths that masquerade as history. Contrary to popular belief, for example, the post-war efforts of Wernher von Braun's German V-2 rocket team in the United States did not provide the crucial military technology employed in Titan and Minuteman intercontinental ballistic missiles. Instead, their efforts resulted in much of the technology for the remarkable Saturn heavy lift space launch vehicles that in the 1960s made possible Project Apollo. The contemporary American Titan and Minuteman military rockets drew on indigenous technology, especially the storable liquid-propellants and castable, composite solid-propellants developed during World War II at Caltech's Jet Propulsion Laboratory. That latter effort also allowed the United States to deploy Polaris submarines with solid-propellant ballistic missiles long before any other state.

Readers of the *AAS History Series* should already know this, and a good deal more besides. Though on occasion one must deal with unalloyed hagiography, the IAA history symposia, in particular, have contributed significantly to our understanding of the history of rocketry and astronautics—world-wide. In this volume, the contribution on early research and development at Reaction Motors may well explain how in the United States kerosene first came to replace gasoline and alcohol as a fuel in large, liquid-propellant rockets. No small matter, that change. You will find other contributions to be equally informative, and provocative.

R. Cargill Hall,
Washington, D.C.

Preface

This volume of the American Astronautical Society's history series publishes eighteen of the papers presented at the fifteenth and sixteenth History Symposia of the International Academy of Astronautics (IAA). These symposia took place in conjunction with two congresses of the International Astronautical Federation, the first held in Rome, Italy, in 1981, and the second in Paris, France, in 1982. In most cases, this is the first English-language publication of these papers, although a few have appeared in the pages of *Acta Astronautica* during the decade since they were first written.

The goal of this publication is to record in one place the historical observations about astronautical developments offered by individuals from broad and divergent backgrounds, with differing perspectives on events, and from many nations. At no other time or place, except in these IAA history symposia, does so diverse a mixture of people come together to analyze the history of humanity's efforts to move into space. Some are professionally trained historians, others are engineers and scientists, and many are of differing nationalities and language groups. But they are united in their interest in understanding the evolution of astronautics broadly considered.

The articles in this collection reflect both this cultural pluralism and divergent interests. The chapters are organized into five basic parts, following the pattern of earlier volumes in this series. The individual articles discuss a variety of subjects from plans for nineteenth century spacecraft, to the development of space science, human training and operations in space, an aerospace corporation's history, through several interesting biographical studies. Many contain excellent illustrations and tables, others rely on text alone to communicate the central message. This collection constitutes an incremental addition to the history of aerospace development throughout the world.

When I received the assignment to edit this volume in November 1991, after the project had lain dormant for several years, several key people helped me assemble the articles and prepare this volume for publication. For their many contributions in completing this project I wish expecially to thank R. Cargill Hall, the *AAS History Series* Editor; Frederick C. Durant III and V. N. Sokolsky, the IAA symposia chairs; Frederick I. Ordway III, who provided considerable assistance, expertise, and support; Lee D. Saegessor, who helped track down illustrations and correct inconsistencies; and J. D. Hunley, who contributed valuable advice. In addition to these individuals, I wish to acknowledge the following scholars who aided me in a variety of ways to complete this editorial assignment: Donald R. Baucom, Virginia P. Dawson, Donald C. Elder, Linda

Neumann Ezell, Adam L. Gruen, Thomas Fuller, Richard P. Hallion, Sylvia K. Kraemer, Cathleen S. Lewis, John L. Loos, Howard E. McCurdy, F. Mark McKiernan, John E. Naugle, Å. Ingemar Skoog, John L. Sloop, Joseph N. Tatarewicz, Shirley Thomas, and Craig B. Waff. I also wish to thank the authors of the individual articles for their patience and helpfulness. A special debt is owed to Ellwood Anaheim and his staff for their persistence in preparing this manuscript for publication. Finally, I extend my thanks to the people at Univelt, Incorporated for their work in bringing this book to print.

Roger D. Launius
Washington, D.C.

Contents

Part I

Early Solid-Propellant Rocketry

Chapter 1

On the Design of N. I. Kibal'chich's Flying Machine (on the 100th Year of its Development)[1]

A. M. Vinitsky[2]

In Russia in the middle of the nineteenth century a large number of military rockets with solid-propellant motors were produced and successfully employed. At that time, Russian scientists and inventors suggested the idea of flying machines that used jet propulsion. To them are attributed two designs—one, a lighter-than-air flying machine using the jet reaction of gas or steam as a motive force (Treteskii, Sokovnin); and the other, a heavier-than-air flying machine propelled by the jet action of a burning liquid fuel carried onboard and using the outside air as an oxidizer (Teleshova).

Nikolai Ivanovich Kibal'chich (1853-1881) introduced an entirely new method, unprecedented in other domestic designs, for making a heavier-than-air flying machine that did not require the atmosphere as a support medium or a source of energy. His device was powered by one or two solid-propellant rocket engines with an automatic feed system (using a timing mechanism), which consumed solid-propellant cells that had different sized surface-burning areas to provide variable thrust. During engine operation, the changing area of the solid-propellant's burning surface provided for controlled, variable thrust, from launch through flight, hovering, and a safe, soft landing.

Flight control was to be provided by the rocket engine's inclination in the vertical plane and its cone-shaped rotation around a joint. As an alternative, Kibal'chich suggested using a second solid-propellant engine in the horizontal plane to improve flight-control attitude—an engine that had to be able to rotate around the vertical axis.

---

[1] Presented at the Fifteenth History Symposium of the International Academy of Astronautics, Rome, Italy, 1981.

[2] National Committee for the History of Science and Technology, U.S.S.R. Academy of Sciences, U.S.S.R.

Stability for the flying machine was to be achieved by applying thrust above the center of gravity and along a single axis. In order to insure that burning proceeded from the end face of the solid propellant, the idea arose to place the propellant in reinforced, cylindrical casings. For propellant, Kibal'chich thought to use compressed gunpowder (a mixture of sulfur, saltpeter, and coal) and other slow-burning explosive materials.

When considering Kibal'chich's nineteenth century project on the whole, one must take note of his originality. His flying machine incorporated, among other features: (1) rocket propulsion for a flying machine; (2) control of vehicle attitude through pitch, roll, and yaw; (3) provision of flight stability by the reciprocal location of the vehicle's center of gravity and the rocket motor; (4) varying speed through a change of rocket thrust (altering of the solid-propellant element); and (5) flight return to a soft landing.

Part II

Rocketry and Astronautics: Concepts, Theories, and Analyses

Chapter 2

The Origins of Inertial Navigation in Space[1]

A. Yu. Ishlinsky[2]

Modern inertial navigation is a vivid example of the realization of an idea that at first looked like pure fantasy, but presently it enjoys wide practical application. From a modern technical point of view, the concept of inertial navigation can be summarized as follows.

In some reference frames $\xi^*\eta^*\zeta^*$ can be considered as inertial, in a gravitational field $\overline{G}(\overline{R},t)$ known for any point $\overline{R}$ and every instant t. There is a standard mass of an object moving in this field, and it is possible to measure continuously the components of an external force, applied to the mass, in projections on axes of coordinate trihedron XYZ, realized on the object.

Projections of the absolute angular velocity of the trihedron XYZ on its own axes, measured, for example, by gyroscopic sensors, are known at every moment. If, for the initial moment, the coordinates of the standard mass and the orientation of the trihedron XYZ are known, then it is possible to determine current coordinates of the mass, and hence those of the object in the reference frame $\xi^*\eta^*\zeta^*$, by continuously solving the kinematic Euler-Poisson equation, which determines the current orientation of the trihedron XYZ, and by integrating dynamic Newton equations for the standard mass. At the same time, the velocity of the standard mass relative to the frame $\xi^*\eta^*\zeta^*$ is also obtained. Thus, the complete navigational problem is solved by autonomous means.

General features of the inertial navigation problem are well understood today. Many different types of systems exist to provide guidance for rockets, aircraft, ships, and spacecraft. The routineness with which these inertial navigation systems operate, however, masks the tortuous path leading to the development of these practical systems.

---

[1] Presented at the Sixteenth History Symposium of the International Academy of Astronautics, Paris, France, 1982.

[2] The Institute for Problems in Mechanics, U.S.S.R. Academy of Sciences, Moscow, U.S.S.R.

The pioneers of inertial navigation research—the Americans M. E. Carrie and F. R. Sweeny, and the Russian B. Alecseev—in the first two decades of the twentieth century made significant strides in resolving most of the fundamental problems of inertial navigation.[3] Their work presented a description of the method for solving navigational problems autonomously for the motion along the Earth's surface by means of two free gyroscopes and a pendulum, but it did not use the concept of double integration of accelerations with which subsequent development of the idea is closely connected. The double integration of accelerations for determining the path traversed was apparently first suggested in R. Wussow's patent, claimed in 1905, given in 1906.[4]

In 1911 in Germany, I. M. Boykov recommended measuring an acceleration by a two-coordinate pendulum and integrating it with the use of a clock mechanism with subsequent calculation of the path traversed by the usual technique.[5] He obtained a patent for this invention in 1928.

O. Dahlke's patent (Germany, 1914), concerned itself mainly with the description of a technique for horizontal leveling of accelerometers, and it mentioned the possibility of using the vertical accelerometer for calculating vertical velocity and determining the height of the object by integrating the signal of this accelerometer. He also pointed out the need for excluding gravity acceleration from this signal. This, apparently, was the first statement of the possibility, in principle, of inertial navigation in space.[6]

In the 1920s the concept of double integration of the acceleration signal, as a possible approach to the solution of the inertial navigation problem, attracted the attention of those involved in astronautics in several locations. For instance, a clear description of such a technique for determining velocity and location of rockets in space can be found in a book by Hermann Oberth in 1929. In his system, the stabilization of accelerometers, with respect to stars, was provided by free gyrostats. Moreover, the book explained the need to account for the dependence of Earth's gravity acceleration (which had been artificially excluded from accelerometer signals) on the distance from the center of the planet.[7]

A detailed description of the concept of inertial navigation in space was contained in a book by Robert Esnault-Pelterie published in 1930. He pointed out that gravity acceleration depended on the location of the spacecraft. The acceleration, measurable by mechanical devices on a moving vehicle, Esnault-Pelterie called "sensible," stressing the difference between this acceleration and full acceleration, which included the accelera-

---

[3] M. E. Carrie, U.S. Patent No., 1,253,666 (1903); B. Alecseev, Russian Patent No., 28451, 30.IY.1916 (1911); F. R. Sweeny, *Geographic Position Indicator*. I086246, 3.II.1914.

[4] R. Wussow, *Apparat zur Bestimmung von Geschwindigkeiten und Wegelangen* (Instrument to Determine Velocity and Distance), German Patent no. 179,477 (1905).

[5] I. M. Boykov, "Navigation Mittels Derivators" (Navigation by Means of a Measure of Drift), in *Zeitschrift für Flugtechnik und Motorluftschiffahrt*, issue no. 11, 1911.

[6] O. Dahlke, *Einrichtung zum Messen von Beschleunigungen* (Equipment for Measuring Acceleration, Velocity, and Paths) (Geschwindigkeiten und Wegen).

[7] Hermann Oberth, *Wege zur Raumschiffahrt* (Road to Spaceflight) (Berlin, 1929).

tion due to gravity. He considered different approaches to integrating sensible accelerations by using a heavy mass in viscous liquid, a gyroscopic pendulum or a pendulum clock, with speed depending on the acceleration along the mean orientation of the pendulum.[8]

For the case of a vehicle moving along a line passing through the Earth's center, with fixed orientation in the inertial space, Esnault-Pelterie developed the equation:

$$\frac{d^2r}{dt^2} = \Gamma - G - \frac{\Gamma_1}{M}$$

where r is the distance from the Earth's center, Γ is acceleration provided by propulsion system, Γ_1/M is acceleration due to environment resistance, and G is gravity acceleration, which is $G = g\, R^2/r^2$.

It is interesting that Robert Esnault-Pelterie, apparently in view of a possible time lag in computing devices, formulated a problem of determining gravity acceleration on points of the trajectory with certain time prediction, and illustrated the possibility of obtaining the solution to this problem. After denoting measured "apparent" acceleration inside the vehicle by:

$$\Gamma - \frac{\Gamma_1}{M} = \gamma$$

Robert Esnault-Pelterie had written an equation for r as:

$$\left(\frac{d^2r}{dt^2}\right)_{r+\Delta r} = \gamma_{r+\Delta r} - gR^2\left(\frac{1}{r^2} - \frac{2}{r^3}\frac{dr}{dt}\Delta t\right)$$

It is now clear that progress in computers made this foresight by Esnault-Pelterie unnecessary. It can be shown by a simple analysis that the inertial navigation system, considered by Esnault-Pelterie, is unstable. An error G of determined distance r is governed by the equation

$$d^2\sigma/dt^2 - gR^2\,(2r + \sigma)\,\sigma \,/\, r^2\,(r + \sigma)^2 = 0$$

where r = r (t) is a true current distance between the vehicle and Earth's center. For the case r = R = const it is possible to write the solution:

$$\sigma(t) = \sigma_0 \,\mathrm{ch}\,(\nu t\,\sqrt{2}) + (\sigma_0/\nu\,\sqrt{2})\,\mathrm{sh}\,(\nu t\,\sqrt{2})$$

where $\nu=\sqrt{g/R}$ is the Schuler frequency and σ, σ_0 errors of initial position and velocity, respectively, are used as input data in the integrating unit.

[8] Robert Esnault-Pelterie., *L'astronautique* (Astronautics) (Paris, 1930).

In 1934 the concept of inertial navigation in space was realized in practice at the Central Aero- and Hydrodynamical Institute by V. S. Vedrov and his colleagues, who developed a technique for the analysis of aircraft motion during a non-stationary spin. Test aircraft were equipped with three accelerometers (g-sensors) and three single-axis gyroscopic angular velocity sensors. The measured g's and angular velocities were recorded by XY plotters.

Graphs of three angular velocity vector components, and of three g-vector components recorded during the flight, were used for calculating time-dependent coefficients in Euler-Poisson equations, for time evolution of the aircraft orientation, and for calculating right-hand-sides of Newton equations, which govern space motion of the point where sensors are installed. Numerical solution to the full system of equations restructured the whole pattern of aircraft motion—its trajectory as well as fuselage orientation and velocities in every point of the trajectory. Thus the full problem of inertial navigation in space was solved. According to modern terminology, it was a typical inertial navigation system without a gimbal, with only one special feature: flight data were processed not by computer but by a group of analysts.

The paper by V. S. Vedrov and his colleagues in 1935 contained equations of an ideally operating system for inertial navigation in space for the case of arbitrary rotation of the frame, connected with sensitivity axes newtonometers (accelerometers); the form of these equations was quite similar to those used presently. They had picked out this group of Euler-Poisson equations:

$$d\varphi/dt = q \sin \vartheta + r \cos \vartheta$$
$$d\vartheta/dt = p - (q \cos \vartheta - r \sin \vartheta) \, tg \, \varphi$$
$$d\psi/dt = (q \cos \vartheta - r \sin \vartheta) / \cos \varphi$$

where φ, ϑ, ψ were attitude angles of aircraft, p, q, r - angular velocities, recorded by XY -plotter.

Expressions were written down for the apparent newtonometers (g,s) measures by newtonometers along axes ξ, η, ζ, connected with aircraft;

$$n_\xi - (g_\xi - j_\xi)/g \qquad n_\eta - (g_\eta - j_\eta)/g \qquad n_\zeta - (g_\zeta - j_\zeta)/g$$
$$g_\xi -- g \sin \varphi \qquad g_\eta -- g \cos \varphi \cos \vartheta \qquad g_\zeta -+ g \cos \varphi \sin \vartheta$$

Here g is gravity acceleration and j_ξ, j_η, j_ζ - projections of the full acceleration vector.

The set of Newton equations is written in this form:

$$du/dt = ru - qw - g\,(n_\xi + \sin \varphi)$$
$$dv/dt = pw - ru - g\,(n_\eta + \cos \varphi \cos \vartheta)$$
$$dw/dt = qu - pv - g\,(n_\zeta - \cos \varphi \sin \vartheta)$$

where u, v, w are the projections of the full velocity vector on the axes connected with an aircraft.

It is stated in the paper that the coordinates of an aircraft in space are obtained by direct integration of the velocity vector after reprojecting it on the immovable frame axes. As an illustration, a typical trajectory of aircraft going out of spin (in three projections) is presented; this trajectory was calculated by the above described technique. V. S. Vedrov's paper seems to contain the most complete of the earliest accounts of the theory of inertial navigation in space.

Chapter 3

Evolution of Space Fiction in Film[1]

Frederick I. Ordway, III[2]

More than eight decades have now passed since the appearance in France of the first film dealing with space travel: Georges Méliès' 1902 production, *Un voyage dans la lune* (A Trip to the Moon). Incorporating elements from H. G. Well's novel, *The First Men on the Moon* (1901) and the earlier Jules Verne classic, *De la terre à la lune* (From the Earth to the Moon) (1866), it became a classic almost immediately. So great was the interest in this film production that even Jules Verne, by this time elderly and ill, visited the set during filming on at least one occasion.

Nothing of distinction followed the Méliès film for more than 25 years. Then, in 1929, Germany's Fritz Lang released his *Frau im Mond* (*The Girl in the Moon*), a full-length silent feature. In quite different ways, both Méliès and Lang dealt with the potential of an age-old dream of landing on, and exploring, our neighboring world in space. The same theme formed a subplot to the H. G. Wells-scripted, Alexander Korda-released 1936 classic *Things to Come*.

These pre-World War II pictures gave rise, during the post-war years, to a seemingly unending array of mostly grade B films about flying saucers, creatures from outer space, and other nonsense. Fortunately, they were punctuated from time to time by such landmark productions as *Destination Moon* (1950) and *2001: A Space Odyssey* (1968). Both films broke ground for their genre, were commercial and artistic successes, boasted magnificent visual displays and special effects, and offered interesting, though probably not brilliant, storylines.

---

[1] Presented at the Sixteenth History Symposium of the International Academy of Astronautics, Paris, France, 1982.

[2] Alabama Space and Rocket Center, Tranquility Base, Huntsville, Alabama, U.S.A.

Nine years after *2001*, another cinematographic milestone was reached with the appearance of George Lucas' *Star Wars*. Splendid and full of vitality, it capitalized on incredible post-*2001* advances in special effects wizardry and a thoroughly entertaining good-guys versus bad-guys screenplay. In the summer of 1980 it was followed by its companion episode *The Empire Strikes Back*. By mid-1981, ticket sales for the two pictures had grossed an estimated $800 million, and *Fortune* magazine estimated that as of October 1980 Lucas' personal fortune derived from his twin efforts had already exceeded $100 million.[3]

The phenomenal success of *Star Wars* inspired other such big-budget, impressive special effects space operas as *Moonraker* (1979); *Alien* (1980); *The Black Hole* (1979); *Close Encounters of the Third Kind* (1977); *Star Trek* (1981) (the motion picture version of the television series); *Outland* (1984); four *Superman* dazzlers; and, more recently, *Star Trek: The Wrath of Khan*, and *Return of the Jedi*.[4]

Films of the 1920s and 1930s

Voyage dans la lune

The Georges Méliès silent production was short, 13 minutes long, and contained elements from the Jules Verne novel *De la terre à la lune* and H. G. Wells' *First Men in the Moon*.[5] From the former came the space cannon; from the latter, the landing on the Moon. Rather incomprehensible to us today was the use of Folies Bergère dancers to insert the spaceship into its launch cannon. However, Méliès apparently looked upon *Le voyage dans la lune* more as a filmed theatrical presentation than as a movie in the sense we understand the medium (Figure 1).

---

[3] Stratford P. Sherman, "The Empire Pays Off," *Fortune*, October 1980, p. 52.

[4] General studies of science fiction films can be found in Brian Ash, ed., *The Visual Encyclopedia of science Fiction* (New York: Harmony Books, 1977); John Brosnan, *Future Tense: The Cinema of Science Fiction* (New York: St. Martin's, 1969); Edward Edelson, *Visions of Tomorrow: Great Science Fiction from the Movies* (Garden City: NY: Doubleday, 1975); James Gunn, *Alternate Worlds: The Illustrated History of Science Fiction* (Englewood Cliffs, NJ: Prentice-Hall, 1975); Walt Lee, *Reference Guide to Fantastic Films: Science Fiction, Fantasy, and Horror* (Los Angles, CA: Chelsea Lee Books, 1972-1974), three volumes; Douglas Manville and R. Reginald, *Things to Come: An Illustrated History of the Science Fiction Film* (New York: Times Books, 1977); Richard Mezer, *The World of Fantasy Films* (New York: A. S. Barnes, 1980); Ed Naha, *The Science Fictionary* (New York: Wideview Books, 1980); Frederick Pohl and Frederick Pohl IV, *Science Fiction Studies in Film* (New York: Ace Books, 1981); Franz Rottensteiner, *The Science Fiction Book: An Illustrated History* (New York: Seabury Press, 1975); John Baxter, *Science Fiction in the Cinema* (New York: A. S. Barnes, 1970); Jeff Rovin, *A Pictorial History of Science Fiction Films* (Secaucus, NJ: Citadel Press, 1975); Chris Steinbrunner and Burt Goldblatt, *Cinema of the Fantastic* (New York: Saturday Review Press, 1972); Philip Strick, *Science Fiction Movies* (London: Octopus, 1976); Jean-Claude Suares, Richard Siegel, and David Owen, *Fantastic Planets* (Danbury, NH: Addison House, 1979).

[5] Jules Verne, *De la terre à la lune* (Paris: J. Hetzel, 1866); H. G. Wells, *The First Men in the Moon* (London: George Newness, 1901).

Figure 1 George Méliès' spacecraft impacts on a protesting Moon in *Le voyage dans la lune* (1902). Courtesy of author.

Born on 8 December 1861, Méliès occupied himself in Paris first as a magician and then in the theater. Later, the emerging art of cinematography caught his attention, and it was not long before he had pioneered the technique of special effects. By 1896, he was making short films based to a considerable extent on them. Six years later he produced *Le voyage dans la lune*, in which he humorously depicted the flight of five explorers to the lunar surface, their encounters with aliens, and subsequent return to Earth. Georges Méliès died in Paris in 1938, forgotten in his time, but most certainly not in ours.[6]

Frau im Mond

The Fritz Lang 1929 classic, *Frau im Mond*, was based on the 1928 novel of the same name by his wife Thea von Harbou.[7] Released in the English-speaking world as

[6] On Méliès see, Georges Sadoul, *Georges Méliès* (Paris: Seghers, 1961), and Pierre Versins, *Encyclopedie de l'Utopie des Voyages Estraordinaires et de la Science-Fiction* (Lausanne: L'Age d'Homme, 1972); Paul Hammond, *Marvelous Méliès* (London: Gordon Fraser, 1971).

[7] Thea von Harbou, Frau im Mond (Berlin: A. Scherl, 1928).

The Girl in the Moon and as *By Rocket to the Moon*, Frau was the first space fiction motion picture to make an effort to assure technical accuracy. It employed as source material Hermann Oberth's treatise *Die Rakete zu den Planetenräumen* (The Rocket into Planetary Space),[8] as well as the works of popular science writer and amateur rocket experimenter, Willy Ley.[9]

Lang hired Oberth as a special consultant to help assure accuracy and realism for his film—within reason, of course. To Oberth's chagrin, his advice was not always taken. The prominent film director would from time to time hide behind what he called "poetic license" and allow, for example, his lunar explorers to walk around the airless world without protective spacesuits. Anyway, Lang argued, H. G. Wells had written in *First Men in the Moon* that there was a thin, Earth-like atmosphere in the deep lunar valleys and other low areas, and he could justify his decision with that evidence[10] (see Figures 2 and 3).

What Lang really wanted from Oberth was a publicity rocket, one that would be fired on 15 October 1929, the day planned for the film's premiere. That proposition suited the retiring Transylvanian scholar who was anxious to experiment physically with some of the ideas he had advanced in his book. Lang managed to get modest financial support from a reluctant UFA (Universum-Film-Aktiengesellschaft) film studio and then added a bit of his own. Despite many efforts, Oberth the theoretician was no engineer; he failed to deliver a rocket for the premiere. Disappointed, he returned home.[11]

Frau im Mond's storyline is engaging. A spaceship is built to travel to the Moon in search of the gold that professor-inventor George Mansfeldt feels certain is there. Lacking funds, his wealthy young assistant Wolf Helius secures help from a group of businessmen represented in the film by the character of Walt Turner. Helius had already carried out considerable research in rocketry, even sending an unmanned spacecraft around the Moon to take photographs.

---

[8] Hermann Oberth, *Die Rakete zu den Planetenräumen* (Munich: R. Oldenbourg, 1923). This book was reprinted in 1925. Both editions were the same, with 92 pages of text.

[9] Willy Ley, *Die Fahrt ins Weltall* (Leipzig: Hachmeister & Thal, 1926), and his edited work, *Die Möglichkeit der Weltraumfahrt* (Leipzig: Hachmeister & Thal, 1928).

[10] Wells wrote, "at the touch of dawn a reek of grey vapour poured upward from the crater floor, whirls and puffs and drifting wraiths of grey; thicker and broader and denser, until at last the whole westward plain was steaming like a wet handkerchief held before the fire; and the westward cliffs were no more than a refracted glare beyond."
"It is air . . . . It must be air—or it would not rise like this—at the mere touch of a sunbeam" (Wells, *First Men on the Moon*, pp. 78-79). Wells continued his tale of the lunar dawn: "Imagine it! Imagine that dawn! The resurrection of the soil, and then this silent uprising of vegetation, this unearthly ascent of freshness and spikes" (*Ibid.*, pp. 90-91). No wonder Lang allowed his travelers to wander around unprotected on the lunar surface, Oberth's objections notwithstanding.

[11] The travails of Oberth at UFA are described by Willy Ley in his history, *Rockets, Missiles, and Men in Space* (New York: Viking, 1968), pp. 114-120, and by H. B. Walters, *Hermann Oberth: Father of Space Travel* (New York: Macmillan, 1962), Chapter 9. See also *The Girl in the Moon* (1927-1929) pp. 63-69.

Figure 2 Construction of the rocket set for *Frau im Mond* (1929). Courtesy of Author.

As the film progresses the spaceship is readied for flight and it blasts off impressively. Aboard are Mansfeldt, Helius, engineer and ship's designer Hans Windegger, his fiancee astronomy student Friede Welten, Turner, and a boy stowaway named Gustav.

Once on the Moon, a greedy Turner tries to take over the spaceship and return to Earth alone, but he is killed in an ensuing struggle. In the fight, however, a stray bullet pierces the oxygen tank, leading to the loss of some of the precious gas. Meanwhile, old Professor Mansfeldt has wandered off and dies falling into a cave-like depression.

Since there is not enough oxygen aboard the spaceship for the return of all the passengers to Earth—even though two are already dead—Helius volunteers to stay on the Moon. As he watches the ship depart for home, he discovers that Friede—with whom he has fallen in love and she with him—has remained behind too.

The visual effects, developed by Konstantin Tschetwerikoff, are quite good as far as the spaceship, the launch and the crowds surrounding it during takeoff are concerned. The Moon's surface fares less well. And we can only grin as the lunar explorers walk normally under 1/6 gravity, clad in rather plain clothes, with no suggestion whatsoever that the Moon lacks an atmosphere. Lang insisted on air. "I cannot have whole scenes played in diving suits," he is reported to have complained.[12]

[12] On Lang see, Paul M. Jensen, *The Cinema of Fritz Lang* (New York: A. S. Barnes, 1969).

Figure 3 An interior scene of the spaceship on route to the Moon, showing the effect of weightlessness, in *Frau im Mond.* Courtesy Cinematheque Francaise.

Just Imagine

What does one do to top *Frau im Mond*, spacesuits or no spacesuits? Well, you might try a musical. That is what director David Butler did in 1930 with *Just Imagine*. People have numbers, not names, in the New York setting of 1980. Butler did not quite hit the mark, though he would have nodded knowingly at our Social Security numbers, postal zip codes, credit card numbers, and telephone area codes (Figure 4).

Anyway, scientist Z-4 built an interplanetary rocket and was bent on flying to the planet Mars. Pilot J-21 signed on to make the voyage to prove himself worthy of marriage to LN-18. On Mars, he and his companions find a primitive human society governed by one Queen Loo-loo. After a series of adventures on the red planet, the crew managed to escape back to Earth, where a court concluded that J-21 and LN-18 can marry. A live-happily-ever-after ending left the audience contented and paved the way for a much more significant endeavor several years later.

Figure 4 A scene from *Just Imagine* (1930) with John Garrick and Maureen O'Sullivan. The film was set in New York City in 1980 but included a flight to the Moon. Courtesy Arfor Picture Archives.

Things to Come

Alexander Korda's film version of H. G. Wells' novel, *The Shape of Things to Come*—the author himself wrote the script—was first released in 1936.[13] Towards the end of the picture in Everytown 2036, John Cabal's grandson, Oswald Cabal, (both roles superbly played by Raymond Massey) organized a trip to the Moon. A sculptor, Theotocopulos, interpreted by Cedric Hardwicke, tried to thwart scientific progress by inciting

[13] H. G. Wells, *The Shape of Things to Come* (London: Hutchinson, 1933), and H. G. Wells, *The Ultimate Revolution* (New York: Macmillan, 1933). See also his *Things to Come: A Film* (New York: Macmillan, 1935, reprinted by Boston, Gregg Press, 1975).

a mob to destroy the giant space gun that launched capsules to the moon.[14] But Cabal, and reason, prevailed and a capsule with Cabal's daughter and young man aboard (King Passworthy IV's son) was fired to the Moon.

Figure 5 Raymond Passworthy, played by Edward Chapman, listens to John Cabal (Raymond Massey, standing) discussing the Moon flight. Courtesy Arfor Picture Archives.

The final scene is unforgettable, with Cabal and Passworthy witnessing the space capsule as it soars Moonward. "There—there they go," exclaimed Cabal.

"That faint gleam of light."

"I feel that what we've done is monstrous."

"What they've done is magnificent."

"Will they come back?"

"Yes. Yes. And go again and again, until a landing is made and the Moon is conquered. This is only a beginning."

"If they don't come back—my son and your daughter—what of that, Cabal?"

"Then presently, others will go."

"Oh, God, is there ever to be any age of happiness? Is there never to be any rest?"

"Rest enough for the individual man—too much, and too soon, and we call it Death. But for Man no rest and no ending. He must go on, conquest beyond conquest. First, this little planet with its winds and ways, and then all the laws of mind and matter

[14] Already outdated—the great space pioneers Konstantin E. Tsiolkovsky of Russia, Robert H. Goddard of the United States, Hermann Oberth of Germany, and Robert Esnault-Pelterie of France all postdate Verne—the cannon persisted in film as the propulsion system for Moon travel. Some of the special effects in this film are memorable, especially the mob converging on the giant cannon.

that restrain him. Then the planets around him and at last out across immensity to the stars. And when he has conquered all the deeps of space and all the mysteries of time, still he will be beginning."

"But . . . We're such little creatures. Poor humanity's so fragile, so weak. Little . . . little animals."

"Little animals. If we're no more than animals we must snatch each little scrap of happiness and live and suffer and pass, mattering no more than all the other animals do or have done," (Cabal points to the image of space in the mirror.) "It is this—or that: all the universe or nothingness. Which shall it be, Passworthy? Which shall it be?" (see Figure 5).

The Serials

These kinds of questions were neither asked nor answered in the swashbuckling, all fun Flash Gordon and Buck Rogers serials that thrilled millions from the mid-1930s to the early 1940s.[15] It was Flash against Ming the Merciless. Buck against Killer Kane; whatever the cost, whatever the means employed, right had to triumph over evil. Scant thought was given to the scientific and technical underpinnings (or lack thereof) implicit in the action (see Figures 6 and 7).

Ten years and a World War later, a new attitude would emerge.

Figure 6 Emperor Ming on his throne in the Flash Gordon Serial, 1936, set on the planet Mingo. Blond Dale Arden, played by Jean Rogers, at left, and Flash Gordon, Buster Crabbe, at right. Courtesy of author.

[15] A. Raymond, *Flash Gordon* (New York: Nostalgia Press, 1967), and P. Nowlan, *Collected Works of Buck Rogers in the 25th Century* (New York: Chelsea House, 1969).

Figure 7 A scene from a 1939 Buck Rogers serial with Constance Moore and Anthony Warde, who played Killer Kane. Courtesy of author.

Films of the 1950s

The year was 1950. The American public had become vaguely aware that flight into space was a possibility. A remote one, perhaps, but something that just might occur in the lifetimes of many then alive. A number of actors were helping to build this awareness.

For one thing, Willy Ley's *The Conquest of Space*, featuring stunning paintings by artist Chesley Bonestell, had come out the year before and had enjoyed a large readership.[16] Also, German wartime V-2 missiles brought to the U.S. and suitably modified by Wernher von Braun and his German-American team of specialists had for some years been probing the upper atmosphere and borders of space. And there were stirrings in the scientific community on extensions of V-2 technology to carry out more ambitious explorations beyond the space frontier.

---

[16] Willy Ley and Chesley Bonestell, *The Conquest of Space* (New York: Viking, 1949).

Destination Moon

Film producer-director George Pal, a master of special effects, sensed that the time was ripe for a realistic approach to space travel in feature films.[17] Specifically, he wanted to depict the first manned flight to the Moon. And he wanted to do it as accurately as astronomical knowledge and the state of the cinematographic art would permit. He would call his creation *Destination Moon*, and would base it on science fiction writer Robert A. Heinlein's 1947 juvenile novel *Rocket Ship Galileo*.[18]

Early in the planning stage, Pal emphasized technical and scientific realism to all involved. He brought in Heinlein and Chesley Bonestell as consultants, who incessantly pep-talked director Irving Pichel and his entire crew. As Heinlein explained, "Mr. Pichel is not a scientist, but he is intelligent and honest. He believed what Mr. Bonestell and I told him and saw to it that what went on the screen was as accurate as budget and ingenuity would permit."[19]

For the time (1950), Lee Zavitz's special effects coupled to Bonestell's lunar backdrops were spectacular (Figure 8). They still are quite convincing. Of course, changes would be in order if the movie were being made today; after all, twelve Apollo astronauts landed there between 1969 and 1972. The mudflat cracks, for example, would be replaced by a fragmented, colorless surface rubble or "soil." And the background mountains would be less jagged and more rounded. But these are details and should not detract from *Destination Moon's* excellence.

Other Films of the 1950s

Pal followed up his success the next year by adapting to the motion picture medium Edwin Balmer and Philip Wylie's novel *When Worlds Collide*.[20] Hurtling towards Earth are the worlds Bellus and Zyra, the former destined to destroy our planet and the other to take its place in orbit around the Sun. A private effort is organized to build a space ark, which is to carry a select group of 40 people to the distant but supposedly habitable planet of Zyra.

Again Pal's set director was Chesley Bonestell, who designed and built a convincing ark (Figure 9). In the film the 40 passengers board and make ready to travel from a doomed world to one they hope will prove suitable. The ramp-launch, created by special effects director, Gordon Jennings, is successful, as is the voyage to Zyra. Meanwhile,

---

[17] On Pal's career see, Gail Morgan Hickman, *The Films of George Pal* (South Berwick, NJ: A. S. Barnes, 1977).

[18] Robert A. Heinlein, *Rocket Ship Galileo* (New York: Scribner's, 1947).

[19] Robert A. Heinlein, "Shooting Destination Moon," *Astounding Science Fiction*, July 1950, p. 6. This story was reprinted with notes as "The Making of Destination Moon," *Starlog*, No. 6, June 1977, p. 19. See also Robert A. Heinlein, *Destination Moon* (Boston: Gregg Press, 1979) which contains "Destination Moon" originally published in *Short Stories Magazine*, September 1950, and "Shooting Destination Moon."

[20] Edwin Balmer and Philip Wylie, *When Worlds Collide* (New York: Frederick A. Stokes, 1934).

Bellus' tidal forces wreak havoc on Earth and its man-made structures begin to crumble under the camera's eye.

Another memorable film came out that same year (1951), *The Day the Earth Stood Still*, directed by Robert Wise.[21] The benevolent alien Klaatu warns the Earth to shape up and control its aggressiveness by disarming. If not, we are to be destroyed by the higher civilization he represents. Just to make sure we comply, before departing Klaatu instructs his faithful robot Gort to keep a watchful eye on us.

Figure 8 Two astronauts land on the Moon in *Destination Moon* (1950), the George Pal special effects extravaganza. Shown here organizing scientific equipment brought with them, the film had a very authentic look to it. The principal problem being the lunar cracks shown here. Courtesy of author.

A couple of years later, George Pal returned to the science fiction scene with H. G. Wells' *The War of the Worlds*.[22] Perhaps unfortunately, the locale was changed from southern England to southern California. The special effects, however, saved the day; they were stunning and gained the film an Oscar (Figure 10). Coincidentally, this film came out at the same time as the silly grade B juvenile film *Invaders from Mars* (Figure 11).

[21] Based on the novella by H. Bates, "Farewell to the Master," *Astounding Science Fiction*, October 1940, p. 58ff.

[22] H. G. Wells, *The War of the Worlds* (London: William Heinemann, 1898).

Figure 9 The Space Ark under construction in *When World's Collide* (1951). Courtesy of Arfor Picture Archives.

Figure 10 In this scene from *War of the Worlds* (1953), American military forces battle three Martian spacecraft. Courtesy of author.

Figure 11 *Invaders from Mars* (1953) was intended to be a 3-D film, but the idea was abandoned during shooting. Courtesy of author.

Space fiction films continued to be a favorite subject during the remaining years of the 1950 decade. Perhaps the most memorable production was *Forbidden Planet*, with its extinct Krell superintelligent society, Walter Pidgeon as Doctor Morbius, and his beautiful and innocent daughter, Altaira, played by Anne Francis (Figure 12).[23] And the most disappointing film of the period? Many would point to the tepid production of Jules Verne's *From the Earth to the Moon* starring Joseph Cotten.

---

[23] The novel based on the film was later written by W. J. Stuart, *Forbidden Planet* (New York: Farrar, Straus & Cudahy, 1956).

Figure 12 A flying saucer spacecraft lands on Altair 4 in a scene from *Forbidden Planet* (1956). Courtesy of author.

The Decades of the 1960s and 1970s

The 1960s brought forth Charles Schneer's nicely packaged version of H. G. Wells' *First Men in the Moon* (1964) and the Stanley Kubrick-Arthur C. Clarke blockbuster *2001: A Space Odyssey* (1968).[24] Despite initially mixed reviews and a somewhat tedious—and to many an unintelligible screenplay—*2001* quickly established a formidable reputation for itself and achieved classic status. For nearly a decade, including the

[24] Arthur C. Clarke, *2001: A Space Odyssey* (New York: New American Library, 1968); Arthur C. Clarke, *The Lost Worlds of 2001* (New York: New American Library, 1972); Arthur C. Clarke, "The Sentinel," *Avon Science Fiction and Fantasy Reader*, 1951. Also see Jerome Agel, ed., *The Making of Kubrick's 2001* (New York: New American Library, 1970); and Frederick I. Ordway III, "2001: A Space Odyssey," *Spaceflight*, 12 (March 1970); 110, and Frederick I. Ordway III, "Voyage to Jupiter, 21st Century Style," *Electronic Design*, 1 (4 January 1967). My experiences as scientific advisor and technical consultant on 2001 are described at length in Frederick I. Ordway III, "Space Fiction in Film—2001: A Space Odyssey in Retrospect," in Eugene M. Emme, ed., *Science Fiction and Space Futures: Past and Present*, Vol. 5, *AAS History Series* (San Diego, CA: Univelt, 1982).

entire period of the Apollo lunar landings and the Skylab orbital space station missions, it remained unchallenged as a science fiction film.[25]

Discovery's mission to the Jovian system is the heart of *2001: A Space Odyssey*. Two astronauts, played by Keir Dullea and Gary Lockwood, and an advanced, vocal input-output computer named HAL (for HAL 9000 series) are the three conscious entities on board. Unconscious, because they are under artificially-induced hibernation, are three other astronauts, who are not to be awakened until the spaceship nears its destination. But, since they are scientists-astronauts rather than flight-astronauts, they are not needed to operate the *Discovery* en route and hence are placed into the hibernation state to conserve air, food, and other life support supplies. When the ship reaches the Jovian system, the awakened astronauts are to conduct whatever investigations may be necessary in an attempt to determine the correlation between the TMA-1 site on the Moon and the giant planet, its orbiting satellites or anything else that may turn up.

The first challenge to *2001* came in 1977 in the form of George Lucas' film *Star Wars*.[26] On a colossal scale, Lucas aligns on one side the Federation of Planets under an evil Emperor and, on the other, the Rebels, one of whose leaders, Princess Leia Organa (Carrie Fisher) has been captured. But she manages to place vital information on the enemy into the memory banks of faithful robots, R2-D2 and C3PO. They, in turn, transfer it to Obi-Wan Kenobi, played by Sir Alec Guiness, who, with Captain Han Solo, owner-captain of the space freighter *Millennium Falcon*, his copilot Chewbacca, Luke Skywalker, and the robots, set off for the rescue and confrontation with the Imperial forces aboard the gigantic space "Death Star" including the Governor of the Imperial Outland Regions, Grand Moff Tarkin; the evil Dark Lord of the Sith, Darth Vader; and hundreds of stormtroopers.

The action builds to the climax. And to an incredible commercial success. Not surprisingly, *Star Wars* paved the way for such competitors as Ridley Scott's space gothic horror film *Alien*.[27] Walt Disney Studios' *Black Hole*, a motion picture version of the television serial *Star Trek*, and *Outland*, a story set in a mining camp on Jupiter's moon Io, were all visual and commercial successes. Even Carl Sagan's television documentary science series *Cosmos* featured a "spaceship of the mind," that was no doubt inspired by goings on in the space fiction film world.[28]

Three years after *Star Wars*, the second of a reported nine episodes appeared, *The Empire Strikes Back* (Figure 13). The cast of familiar characters returned, Darth Vader, Princess Leia, robots, and all. And the battle between the forces of good and evil continued. Empire was followed by *Return of the Jedi* in 1983.

From Georges Méliès to George Lucas and his contemporaries, four generations of film goers watched all manner of amazing things happen before their eyes, from a rocket lodged in the Man in the Moon's right eye to super-maneuverable craft engaged in deep

---

[25] On Kubrick see Norman Kagan, *The Cinema of Stanley Kubrick* (New York: Holt, 1972).

[26] George Lucas, *Star Wars* (New York: Ballantine Books, 1976). See also *The Star Wars Album* (New York: Ballantine Books, 1977).

[27] R. J. Anobile, ed., *Alien*, screenplay by Dan O'Bannon (New York: Avon, 1979).

[28] Carl Sagan, *Cosmos* (New York: Random House, 1980).

space dog fights. Some spectators sought realism, demanding that storylines and sets alike be convincing and not too far removed from the technology of today. Others, be they from the Flash Gordon-Buck Rodgers generation or that more familiar with *Star Wars* and *Star Trek*, were attracted by high adventure and by situations bordering on the impossible. They wanted to escape from reality rather than to meet it at every turn.

Figure 13 The spacecraft Millennium Falcon is being prepared for takeoff in a scene from *The Empire Strikes Back* (1980).

Conclusion

And so we have reached the 80th anniversary of the Georges Méliès *Un voyage dans la lune*. Cinema theaters everywhere are inundated with all manner of science fiction films, some dealing with space, some with other subjects. Steven Spielberg's *E.T.—the Extra-Terrestrial*, *Poltergeist*, the Disney *Tron*, the sequels to the first *Star Trek* film, *Bladerunner*, the list seems endless. Reminded we may be of the successes of the Space Shuttle, Vikings on Mars, Veneras to Venus, and Pioneers and Voyagers to Jupiter and Saturn, but man still thirsts for what lies beyond. And that is what makes the seemingly impossible become a possibility.

Part III

The Development of Liquid- and Solid-Propellant Rockets, 1880-1945

Chapter 4

Rocketry Personnel Training in the U.S.S.R., 1924-1936[1]

E. K. Moshkin and A. F. Nistratov[2]

In the late 1920s and early 1930s great attention was given in the Soviet Union to the study of interplanetary communication problems. Various working groups and teams were set up by enthusiasts of jet propulsion to discuss the possibilities of interplanetary flight and to develop the schematic diagrams of rockets and their engines. In addition, some scientists devoted much attention to determining optimal parameters of rocket engine combustion processes and to the estimation of the best ballistic paths.

To facilitate the further development of rocketry, senior leaders believed it necessary to centralize the study of jet propulsion on problems under government control. Therefore, it was expedient to train specialists capable of embracing a wide range of rocket flight problems: designers, engineers, and technicians to conduct theoretical and experimental investigations. The result was the creation of an aggressive rocketry training program for technical personnel beginning in the mid-1920s and its expansion during the rest of that decade in concert with the rapid scientific and engineering development in the Soviet Union.

Several significant Soviet scientists were involved in this process. Perhaps the most important among them was K. E. Tsiolkovsky, whose works originated the idea and skillfully and quickly converted most readers into ardent enthusiasts. Additionally, F.A. Tsander, whose theory and application showed the realization of jet propulsion problems, and who was instrumental in developing a viable personnel qualification pro-

---

[1] Presented at the Sixteenth History Symposium of the International Academy of Astronautics, Paris, France, 1982.

[2] National Committee for the History of Science and Technology, U.S.S.R. Academy of Sciences, Moscow, U.S.S.R.

gram. Moreover, B. S. Stechkin, who outlined the principles of jet designing and development, was also involved. B. S. Zemsky, N. I. Dobrotvorsky, N. A. Zhuravchenko, V. P. Vetchinkin, and V. P. Glushko also made important contributions to the progress of rocketry. Among those who joined with this group were experienced organizers and outstanding scientists including V. P Kapersky, M. G. Serebryakov, M. J. Lapirov-Scoblo, M. A. Rezoumov, D. A. Grave, J. I. Perelman, E. O. Paton, N. A. Rynin, B. I. Sreznevsky, K. K. Seminsky, V. I. Shaposknikov, M. S. Belayev, G. A. Polevoy, Z. G. Pyatetsky, O. V. Kholtshev, and others. Without question, the first echelon of enthusiastic organizers of the personnel qualification system was highly impressive and worked hard to assure the success of the training program.

The complex system of training rocketry specialists underwent a number of changes with time. At first there appeared different groups, teams, and societies; papers and talks were given and exhibitions and conferences were held. These efforts made an invaluable contribution to the development and progress of the theory and application of rocketry. In the 1930s, rocketry personnel training was conducted under the auspices of the Central Council of the Osoaviakhim Society, training that became systematic and planned: Group for the Study of Jet Propulsion (GIRD) was established, which, in fact, became the first Soviet Higher School of Rocketry.

We should like to dwell first on the activities of the scientific communities which were engaged in jet propulsion prior to the establishment of the institute. The first conspicuous steps were made in 1924, when, at the January session of the theoretical section of the Moscow Society of Amateur Astronomers, F. A. Tsander gave a paper "On Spacecraft Design and Flights to Other Planets." In this lecture he suggested the establishment of a "Society for the investigation of interplanetary communication."

A few months later, in April 1924, in response to Tsander's recommendation, a "Section of Interplanetary Communication" was set up by cadets of the Zhukovsky Air Force Academy. A month later, M. J. Lapirov-Scoblo gave a lecture on interplanetary flight in the Big Hall of the Polytechnic Museum, and new members enrolled in the Society for the Investigation of Interplanetary Communication. In October 1924, the Society held a conference on "Flights to Other Worlds," and, in November, V. P. Vetchinkin presented a lecture on the possibility of interplanetary flight. Various lectures were delivered by enthusiasts and experts at aircraft plants and factories, at the N. E. Bauman Moscow Technical School of Higher Learning as well as in institutions at Leningrad, Kharkov, Saratov, Ryazan, and Toula. In June 1925, D.A. Grave presented his "Appeal to the Societies for Space Exploration and Conquest." Also in June of the same year, the Society set up by D. A. Grave organized an exhibition devoted to the problems of space exploration. In 1927, the Moscow Inventors Association held its "First Exhibition of Models and Designs of Interplanetary Apparatuses and Mechanisms." In February 1929, at the Interplanetary Travel Amateur Society set up at the Leningrad Institute of Railway Engineers, J.I. Perelman presented a comprehensive plan devoted to the development of practical work in jet propulsion.

As a result of these initial activities, by 1930 substantial premises had been laid for subsequent fundamental and experimental investigations and also for training highly qualified specialists, and a groundswell of support for the effort had been organized. In response, the Central Council of the Osoaviakhim Society issued a directive on 14 June

1932, that recognized the vast amount of propaganda and popularization of rocket ships and of the active training of highly qualified personnel.

In September 1932, on the initiative of S. P. Korolev, two "Special Courses" were started at the Central Jet Propulsion Investigation Group: a senior group for training spacecraft designers and a junior group for training technicians. The Special Courses were attended mainly by the engineering staff, junior researchers, and senior students. The scientific background of the audience allowed an up-to-date treatment of the theory and calculation methods related to interplanetary flight. S. P. Korolev himself participated in writing the syllabus for the senior group. The term of study lasted one year and amounted to 475 teaching hours. The object of the course was to qualify both senior and junior designers, and senior and junior engineers. Within a short time, the course of instruction was extended to 1000 hours, which equated to a 2.5 year program. In 1934, the first group of specialists graduated from these courses, and they were employed by various rocket enterprises.

To understand more fully the quality and depth of the program, it would be useful to review one of the senior group syllabuses. In the early 1930s it included these general items:

1. A course on jet propulsion (25 hours) embraced theory and problems of rocket structure and engine design. M. K. Tikhonravov gave lectures on the aims and purposes of reactive propulsion; J. A. Pobedonostsev delivered a course of lectures on the fundamentals of the pressure fed liquid propellant jet engine.
2. Lectures on aerodynamics and aviation theory (40 hours), covered the basic laws and provisions of practical aerodynamics and the problems of aerodynamic flight.
3. Physiology of rocket flight (10 hours).

A specialized part of the overall curriculum for the senior group comprised the following sections:

1. V. P. Glushko lectured on propellants and combustion (30 hours), which embraced propellants for jet engines and basic theories of combustion.
2. M. K. Tikhonravov delivered the majority of lectures on material properties and the impact of technological designing and external factors on structural strength and construction reliability (30 hours). Here, as well as in other courses, for the first time in the history of science, the lecturers treated the fundamentals and principles of classical disciplines in terms of their application in jet propelled aircraft and engines.
3. V. P. Vetchinkin and B. S. Stechkin delivered lectures on the propulsion theory (60 hours), which included the derivation and analysis of the basic equations for variable mass body motions, thermodynamics, heat transfer, and heat calculation methods. Additionally, V. V. Uvarov gave lectures on gas dynamics.
4. A. N. Zhuravchenko gave lectures on experimental aerodynamics and the dynamics of rocket flight (40 hours) involving the derivation of a set of equations for rocket ballistics, the theory of aerodynamic rocket flight, the laws of gravity and air drag, and rocket control principles. A. N. Zhuravchenko treated these problems in terms of reactive propulsion.

5. Lectures on instrumentation (15 hours) with special emphasis on instruments for practical use in engine and rocket testing.
6. Lectures on rocket operation (25 hours) formulated the main tactical-technical requirements for rockets and engines and tactical-operational conditions for spacecraft.

Many celebrated professors, such as S. P. Korolev, M. K. Tikhonravov and others attended the lectures with a view to broadening their outlook.

These lectures were followed by practical training (190 hours) in independent aerodynamic, heat and strength calculations, and also in the actual designing of rockets, engines, and engineering units. M. K. Tikhonravov supervised this course. The students attended classes once or twice in a 6-day week from 6 to 10 p.m., after work. An optional course was also included in the syllabus.

In late 1933, GIRD became a part of the newly originated Rocket Research Institute (RNII) and, in accordance with a decision adopted by the Council of Labor and Defense, all the work connected with scientific and technical propaganda, publishing of scientific literature and personnel training, was reorganized and placed under the Reactive Propulsion Group of the Military-Scientific Committee, established in January 1934 in the Central Council of the Osoaviakhim Society of the U.S.S.R.

The rocket group directed by I. A. Merkulov gave a final course and prepared the first graduation group of rocket designers. In 1936, the Special Course was arranged for new groups of students. V. P. Glushko was responsible for the scientific and methodological guidelines for training engineers in rocketry at the Military-Scientific Committee. Unfortunately, few people have preserved synopses of lectures presented by Glushko and others, including M. K. Tikhonravov, N. M. Dobrotvorsky, A. I. Zhouravchenko, B. M. Zemsky, V. P. Vetchinkin, B. S. Stechkin. We have only a small number of historical documents, a few pictures of student groups, some diploma papers, and diploma design themes.

From what remains available, however, we have reconstructed an assignment example. The student in the 1930s was supposed to:

1. Design a meteorological rocket with a rated flight altitude of 10 km, using methane and liquid oxygen as propellants
2. Study the structural materials available;
3. Estimate heat design for jet exhaust;
4. Make complete calculations of the combustion system, its construction, strength, and weight;
5. Determine the rocket center of gravity; and
6. Develop the general design view and prepare the necessary auxiliary drawings.

Designs proposed in the above assignment required substantial knowledge of the subject. Indeed, the knowledge of graduates in the course on rocketry was quite comprehensive, at least profound enough to ensure further success in the activities of GIRD and of RNII.

A review of the diploma designs, and a reading of the lecture synopses, convinces us that many special problems concerning the theory of rockets and engines were, for the first time, developed comprehensively. Some engineering solutions advanced by the students were much ahead of their time. To name but a few, the students designed and developed reusable rocket vehicles, rocket engines with high economic parameters, original stage combinations of ballistic and aerodynamic vehicles, etc. The knowledge acquired during the course enabled many specialists in later years to develop and present their subjects to scholars engaged in rocketry, and to contribute to the training of new generations of engineers for design departments and research institutions engaged in rocketry and astronautics.

From 1935 to 1938, the Reaction Group Section issued three series of publications on "Jet Propulsion" compiled by budding and leading scientists. In later years there appeared textbooks, handbooks, and monographs devoted to problems of jet propulsion. The literature published in this area made a significant contribution to further improvement of personnel training in rocketry in the Soviet Union.

Chapter 5

The Supersonic Wind Tunnel Installations at Peenemünde and Kochel and Their Contributions to the Aerodynamics of Rocket-Powered Vehicles[1]

Rudolph Hermann[2]

With the goal, in 1937, to build an aerodynamic-ballistic research institute to provide all aerodynamic, stability, and heat transfer data needed for the development of rocket-powered vehicles, space vehicle research in Germany moved into a new era. This paper describes the design and construction of the resulting supersonic wind tunnel with a 40 cm x 40 cm test-section, which became operational at Peenemünde in 1939 and was moved to Kochel in 1943. Seven wind tunnel Laval nozzles, ranging in velocity up to Mach 5, were used in this facility and their theory, design, and construction, as well as their wave patterns, occupy an important part of this discussion. In addition, the results of tests in this facility for vehicles without wings, such as the A-4, and for vehicles with wings, including delta wings such as the "Glider A-9" or the "Wasserfall," provide useful case studies of the type of research conducted there. Finally, I will describe the influence of jet exhaust on drag and stability.

[1] Presented at the Fifteenth History Symposium of the International Academy of Astronautics, Rome, Italy, 1981.

[2] Professor Emeritus, Aerospace Science and Engineering, The University of Alabama in Huntsville, Alabama, U.S.A.

Purpose, Design, and Operation of the Supersonic Wind Tunnel Facilities

Purpose and Goal

The goal in 1937 was to build an aerodynamic-ballistic research institute capable of furnishing—in a reasonable time frame—all aerodynamic, stability, aerodynamic control, and heat transfer data needed for the development of numerous projects such as supersonic projectiles (fired from guns), rocket-powered vehicles without wings, those stabilized by fins (called missiles), and rocket-powered supersonic vehicles with wings and fin-assemblies or with delta wings. This required the design and construction of supersonic wind tunnels of the highest obtainable Mach speed (at least 5 and above), which also should operate in the subsonic range, since all rocket-powered vehicles start with zero velocity.

From 1934-1937, at the Aerodynamic Institute of the University of Aachen, under the direction of Professor Carl Wieselberger, the author developed, built, and operated a supersonic wind tunnel of the blow-down type with a 10 cm x 10 cm test section up to M = 3.3 with a 3-component balance. Financed by the German *Luftwaffe*, some of the first models tested were anti-aircraft projectiles.

On 6 January 1936, Wernher von Braun of the German Army Ordnance organization, brought the drawing of a pointed, slender body with fins to be tested for drag and stability up to our maximum Mach speed. It was the rocket-powered missile later designated A-3. The first proposed fin assembly would have caused an unstable missile, but further experiments at the institute produced a more stable arrangement (see below, A-3 Vehicle).

For more detailed measurements, larger models and larger tunnel cross-sections were required. More power was needed because of model size and the requirement to test at higher Mach numbers. This showed the necessity of building such a facility with the aid of the Army Ordnance at the Experiment Station under construction at Peenemünde. Our work started in April 1937 with 20 people, all committed to combining available scientific basic concepts with a detailed engineering design and solid construction of all test equipment and modern measuring devices to facilitate maximum tunnel usage.

Additional assistance for the scientist was provided by a series of laboratories, especially those specializing in optics (Schlieren, Interferometer), thermodynamics of high speed gases, heat transfer, electric controls for operation of powerplant and air drying installation, electronic measurement devices, and physics (X-ray densitometer). These laboratories produced the measuring devices needed to solve the problems arising during the development of new missile projects and their counterparts in the wind tunnel.

From the start, installation planners intended that test goals could be facilitated by having a Chief Engineer, with a staff of qualified designers and with the help of ample design offices, combined with a machine shop under the direction and control of the design staff. The result was that ideas and proposals of scientists could be designed and fabricated properly and made ready for testing with a minimum of red tape.

Design and Operation

The work at Peenemünde started in April 1937 with a staff of 60 people, but not until the summer of 1939 did the tunnel first become operational. The wind tunnel continued in use until the fall of 1943, when, following a succession of Allied air raids, it was moved to Kochel—along with 200 staff members—rebuilt, and put back in operation by the spring of 1944. Step-by-step, thereafter, the Kochel staff improved the tunnel through the fall. In Peenemünde, the vacuum reservoir was 1,000 m^3 (Kochel 750 m^3); six vacuum pumps with 800 KW total (Kochel first four pumps); two test sections of 40 cm x 40 cm (Kochel one test section); one test section of 18 cm x 18 cm, capable of continuous operation through M = 3.3. During blow-down operations, Peenemünde had a running time of 20-25 seconds; while Kochel had only 15 seconds. The range of Mach numbers covered and nozzles are described in Section B.

Figure 1 shows the complete layout. The air entered the entrance cone, in front of which was the air drying filter, not shown. Air passed through a honeycomb, entered the test chamber with two Laval nozzles, producing a parallel free jet, which was captured by the bell mouth of the variable diffuser. The diffuser had a fast moving throat to control the pressure in the chamber needed for a parallel jet. The air flowed through the quick acting valve, which opened and closed at the beginning and end of the usable running time. A fixed diffuser brought the air into the vacuum reservoir.

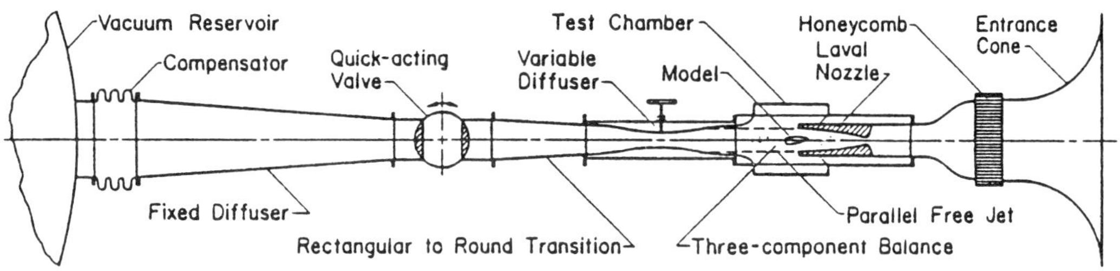

Figure 1 Supersonic wind tunnel layout (40 cm x 40 cm) at Peenemünde (1937).

A photo of the test section installed at Kochel is shown in Figure 2. A "Wasserfall" model is installed, ready for testing, supported by the mount which varies the angle of attack. This could be done from outside by push-button control without opening the test section; another feature for maximum tunnel usage.

An important piece of measuring equipment was the 3-component balance seen in Figure 3, which determined drag life and pitching moment. It was built in Peenemünde in 1939 and also used in Kochel through 1945. It was ruggedly built and installed in a pressure tight chamber, which was evacuated at each run down to the low static pressure of the respective Mach speed. The angle of attack changer can be seen here in the schematic. To determine the Center of Pressure (C.P.), the drag was measured in two components. In customary subsonic wind tunnel balances for aircraft use the lift is measured in two components. This system was needed because many of our models had no lift near zero angle of attack, but they always had drag.

The Schlieren-optical system built in 1937 had two nearly parabolic mirrors, 50 cm diameter each, with 10 m distance. It observed the flow fields around models; large glass walls on both sides of the test chamber (see Figure 2) allowed full view for Schlieren of Interferometer.

The Schlieren system was mounted on a four-wheel carriage running along the test section; and also vertically, to move the system quickly and safely for observation at any point, and also from one test section to the other. The Interferometer (since 1943) also had mirror diameters of 20 cm and a field of view 9 cm x 10 cm.

The installation of the air drying system was completed in the spring of 1940. It was based on detailed experiments on condensation shocks caused by humid air, which were done by the author in Aachen in 1935-1936. The dryness requirement increased with the Mach Number. A two stage drying system, specially developed for intermittent operations, used 600 KW of power for heating and airflow. The dryness reached was 0.2 g Water/kg air, or 0.0002 dimensionless. This was a factor of 10 smaller than that used before (at M = 3). This guaranteed shock free flow to M = 5.2 in the humid climate.

Figure 2 Test section at Kochel in 1945.

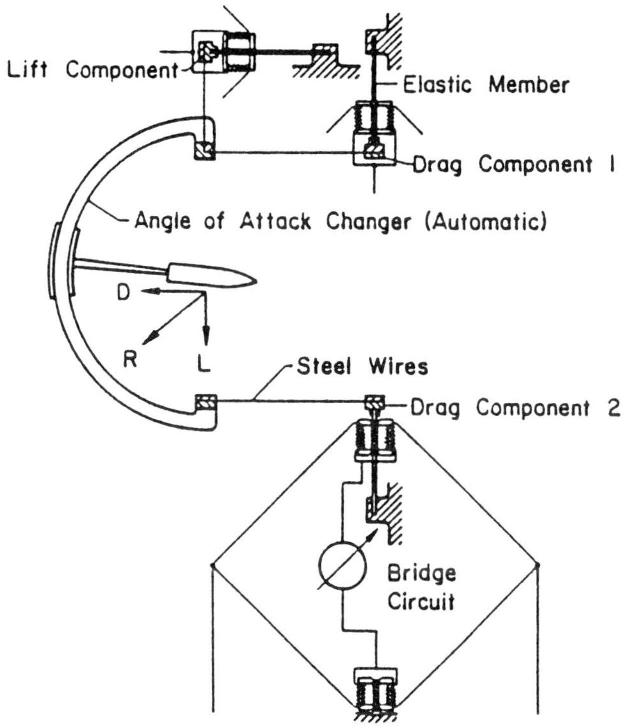

Figure 3 Schematic of the 3-component balance.

Aerodynamic Data for the Supersonic Tunnels and Some Basic Research Conducted

Laval Nozzles, Theory and Design

The nozzle within a supersonic wind tunnel is of paramount importance for conducting correct and meaningful experiments. Essentially, it has two tasks. The first is to produce the desired supersonic velocity in the test chamber. To accomplish this, a Laval type nozzle is applied, which produces subsonic velocity in the converging section, M = 1 at the throat, and supersonic velocity in the divergent section. The ratio of exit area to throat area determines the Mach number. For non-movable nozzles, this requires a set of nozzles with decreasing throat area for increasing the Mach speed, since the exit cross section is kept about equal to accommodate the models.

The second involves the flow of the so-called measuring rhombus, a parallel construction of constant Mach speed and free of shocks. This can be achieved by forming the walls of the nozzle according to the method of characteristics first given by Prandtl and Busemann, applicable to two dimensional flow. Hence, the nozzles are two dimensional, with two equal portions symmetric to the axis. The method is basically graphical, but was improved by the staff of the wind tunnel by substituting calculations.

The early nozzles used at Peenemünde and Kochel had two distinctions. First, they were modeled on the Busemann method, and represented the shortest length of nozzle which could be achieved by any means for a given throat size and shape. Second, their mechanical design and construction was novel, accurate, and effective when boundary layer corrections were needed.

The shortest possible length was important for high Mach—4, 5, 10 (for future projects)—for reasons of weight, handling, and for minimizing boundary layer thickness. Each nozzle had a particular radius of curvature at the throat, producing expansion up to the point of inflection, where the angle, made by the profile with nozzle axis, was denoted by Φ_β. If υ is the Prandtl-Meyer angle which provides the desired Mach number at the exit, then, theory shows, the condition: $\Phi = 1/2\upsilon$ causes the minimum length of the supersonic portion. This condition was satisfied in all our nozzles. Figure 4 illustrates the discussion and shows wave patterns for the nozzle M = 4.4. This high Mach speed was selected to show the complexity of the design (completed in 1938). From a large original drawing it is reduced twice by photographic means. Lower Mach speeds were much simpler, for instance M = 1.56 had only eight wavelets.

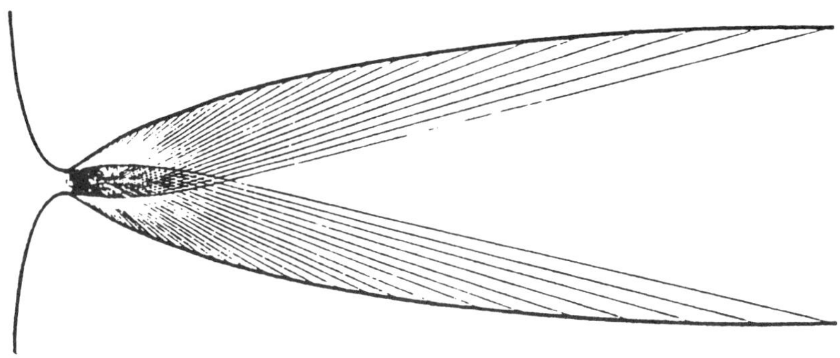

Figure 4 Wave pattern for nozzle M = 4.4.

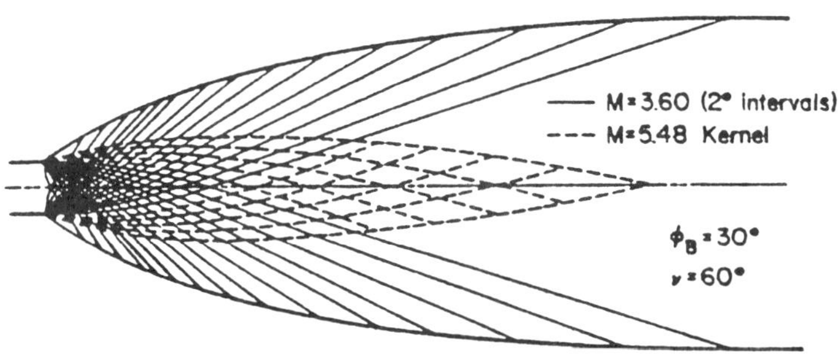

Figure 5 Wave pattern for sharp throat nozzle.

Table 1 shows characteristic dimensions of six of our nozzles (M = 1.22 to 4.38) in use for many years, and for the last (M = 5.18), which was completed in January 1945 but was never thoroughly tested. The full length of the measuring rhombus gave the allowable maximum model length. The last column is the slope of profile at exit (no dimensions). It is equal to the angle of divergence in radians. The divergence is the consequence of making allowance for the displacement thickness of the boundary layer. The particular value of this angle was obtained by experiments.

Table 1
CHARACTERISTIC DIMENSIONS OF WIND TUNNEL NOZZLES
(40 cm x 40 cm)

Design Mach No.	$\Phi_{\beta max}$ degree	Throat width mm.	Exit width mm.	Length of supersonic part mm.	Length of measuring rhombus mm.	Slope of profile at exit radians.
1.22	2	386.16	399.76	271.0	?	0
1.56	~ 6	325.82	403.76	495.8	480	0.008
1.86	~11	259.20	401.84	593.7	602	0.012
2.48	~20	152.10	404.76	787.4	904	0
3.24	~27	71.52	408.74	920.0	1204	0.004
4.38	35	25.94	400.0	1080.3	1704	0.010
5.18	?	14.30	422.56	1224	?	0.010

With the higher Mach speeds, the distance between the throat and the inflection point became physically smaller and smaller, and hence harder to manufacture (and measure) according to design specifications. This led to the idea of a throat with a sharp edge, immediately followed by an expansion around the corner up to $\Phi_{\beta max} = 1/2\upsilon$. The inflection point moved up to the throat, reducing the length, *l*, to an absolute minimum. For M = 5, $l/h = 2.56$; for M = 10, $l/h = 4.98$ (*h* is height of the exit cross section.)

Nozzles with sharp throats can be designed with advantage, since all wavelets originate at the sharp edge. Figure 5 illustrates where a nozzle for M = 3.60 is designed from given wavelet field, called Kernel, corresponding to M = 5.48

Nozzle Construction

The nozzle had a composite construction consisting of a steel base, brass sides or formers, and a cement profile. The cement's name was "monolith." The main structure was nothing more than a mold for the cement. The formers, fixed to the sides of the nozzle structure, were of brass, 5mm thick. They were shaped to the calculated nozzle profile with an accuracy of about 0.01 mm, final shaping being done by hand. Figure 2 shows part of those details. The cement was poured into the nozzle structure and shaped with the aid of a straight edge. The depth of the monolith layer ranged from 10 to 20 mm. Monolith had several advantages over wood or steel: flexibility to changes in the profile, no special machinery required such as for steel, and no deformation or cracking of wood. A complete pair of nozzle blocks required a total of about 2,000 work-hours to construct.

Experimental Investigation of Nozzle Flow

Once the nozzle blocks were completed, the flow field was carefully investigated. The procedure varied somewhat with the Mach speed, but consisted of Schlieren photography, static tube, and/or pitot tube measurement. For $M < 3.3$, pressure distribution along the axis was measured with a long static tube, which stretched from upstream of the throat to downstream of the measuring section. The tube had nine static holes, and the tube was moved lengthwise to cover all spots. Mach No. distribution, measured from Schlieren, was good up to $M = 3.3$, above that the definition of the wavelets was poor, even when ruled scratches in the nozzle surface were used. Then the static pressure and Mach number were calculated from a pitot pressure reading alone. The various variables were converted into each other, using isentropic flow, an assumption that was carefully proven previously.

In Figure 6, the first three show both pressure tube and Schlieren determinations separately. The last two nozzles show the mean of pressure tube and Schlieren. The agreement, on the whole, between pressure and Schlieren measurement was surprisingly good. It follows from the figures that the maximum deviations of Mach speed from the mean were about ± 2.5% in the worst case ($M = 1.22$; not shown here) and about ± 0.5% in the best case ($M = 4.38$).

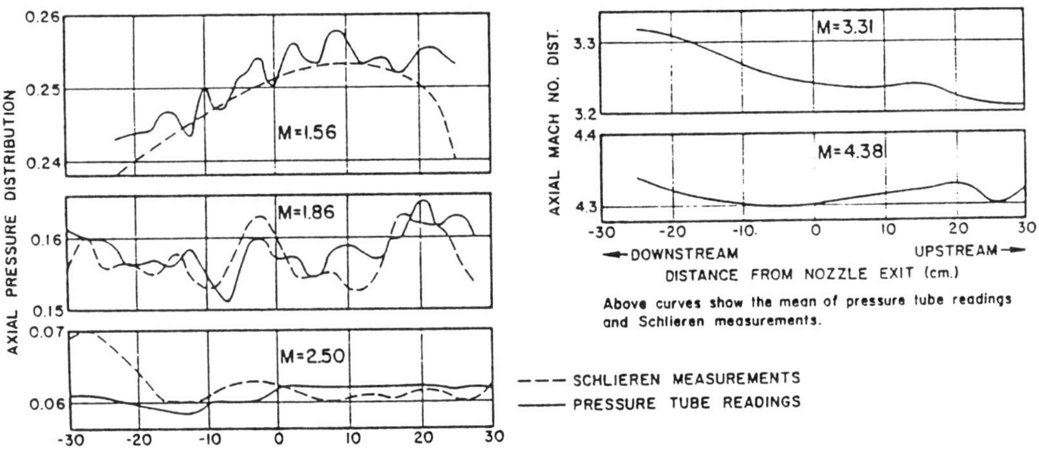

Figure 6 Measured axial distribution for pressure and/or Mach number for 5 nozzles.

Some Basic Research Conducted

When the reader assesses the amount of scientific and engineering accomplishment presented in this paper, it is obvious that this could be accomplished only by a great amount of basic research, conducted simultaneously by the staff of our supersonic wind tunnel installation and by outside consultants, mostly professors, who worked on contract with us. The author, in the summer of 1945, had prepared a summary of basic research in gasdynamics, aerodynamics, thermodynamics, and ballistics, a fraction of which is now used.

Theoretical Gasdynamics

1. Flow in the Laval nozzle and the application of the Prandtl-Busemann method, with corrections to include the boundary layer for the purpose of producing a homogeneous parallel flow.
2. Theoretical work to investigate the stationary, normal shock in the Laval nozzle, including the role of entropy, also the "normal condensation shock."
3. Professor Tollmien of Dresden developed a method of calculating supersonic flow with rotational symmetry. Pressure distribution along the A-4 body was calculated and, together with another method developed by Professor Sauer of Aachen, compared to experiment.
4. The Source-Sink method, first developed by von Kármán and Moore, was used to calculate the pressure distribution for axisymmetric, slender bodies and compared with experiment.
5. The theory for supersonic flow around an infinite wedge and cone was evaluated with the help of special diagrams and nomograms. The shock waves around cones were examined in a series of Schlieren pictures and compared with the theory.

Aerodynamic Performance of Rocket-Powered Vehicles Obtained in the Supersonic Tunnels

Requirements for Rocket-Powered Vehicles

Some requirements for rocket-powered vehicles when under development appear below:

1. Optimum aerodynamic shapes for minimum drag.
2. Lift forces required for certain missions. They are different for a ballistic trajectory, a gliding trajectory or an anti-aircraft missile trajectory.
3. Aerodynamic stability in both subsonic and supersonic ranges. Vehicle is required to be definitely stable, but stability must not be too large since, otherwise, the control requirement of movable rudders with their hinge moments will be excessive.
4. Aerodynamic pressure distribution loading and aerodynamic heating on bodies, fins, wings, rudders—needed for the stress analysis in structural design.
5. Hinge moments for rudders in the air stream or in the exhaust jet, required for control purposes, for pitch, yaw and roll motion.

Two basic groups of vehicles may be distinguished:

1. Rocket-powered vehicles without wings stabilized by aerodynamic fins; typical: A-4, with its forerunners A-3 and A-5.

2. Rocket-powered vehicles with wings, stabilized by a portion of the wing or by extra fins; typical: "Glider A-9" with lift required to stretch the ranges or "Wasserfall" with lift required for maneuverability as anti-aircraft missile.

Rocket-Powered Vehicles Group (1)

A-3 Vehicle

A forerunner of A-4, this craft had a body shape similar to the A-4, but with long, narrow fins with a ring at their end required for containing radio antennas (see Figure 7). Aerodynamic data (three-components) measured by the author in the supersonic wind tunnel at Aachen with 10 cm x 10 cm cross section in 1936 and early 1937. A-3 was the first vehicle, provided by wind tunnel tests, to be fin-stabilized in the supersonic range (up to M = 3.3). It had a slender body of eight caliber length with sharp pointed nose for minimum resistance in supersonic flight. It was fired four times in December 1937, reaching only moderate height (400 m) and subsonic speed. The A-3 was aerodynamically stable around its center of gravity (C.G.) (no "somersaulting"), and its trajectory went into the wind, as required for aerodynamically stable vehicles. The reason for the short flights was that the gyro control system was too slow and the moment produced by the exhaust jet rudders too small to counteract heavy winds (about 25 m/sec).

A-5 Vehicle

The A-5 was also a forerunner of A-4. Due to the trajectory failures encountered with the A-3, another small vehicle was developed with thrust, weight and body size about the same as the A-3, but with a fast control system to avoid the deficiencies of the A-3. Four large air rudders were added at the end of the fins, to produce pitch, yaw and rolling moments to counteract strong natural winds (100 m/sec) at higher altitudes. It was called A-5, since A-4 was already on the drawing board. The fin and rudder assembly was named V12. On purpose, the body of A-5 had the same geometry as that of A-4. Nearly all aerodynamic data gained for A-5-V12 could be used later for the A-4-V12. Starting in October 1939, and continuing through 1941, about 70 to 80 of the A-5 vehicles, with the completely new control system using a stabilized platform, were successfully launched, reaching a ceiling of 8 km, a range of 15 km and a maximum speed about 300 m/sec, close to M = 1.

A-4 Vehicle

Basic Aerodynamic Data. The political name of this rocket was V-2. The configuration of the A-4 can be seen in Figure 8, and the shape as a wind tunnel model can be seen in Figure 9. As outlined above, all aerodynamic data, such as drag, lift and center of pressure (C.P.) and others needed for development were measured. These were obtained in subsonic tunnels up to M = 0.85 and in supersonic tunnels for M = 1.2 and up to M = 4.4. For M = 1, it was not possible to test at this time, since transonic tunnels, in the present sense of that word, did not exist. The center of pressure measurements on both sides of M = 1 showed forward motion, indicating possible instability. Hence, the author suggested, in July 1938, a drop test from airplanes at 7,000 meters, of rocket models about 1.5 meters in length, built from steel. These tests were carried out in 1939 by the Air Force, but under direction of and evaluation by the staff of the Peenemünde supersonic tunnel. The models reached transition through M = 1 at about 1,000 meters in altitude, which could be observed with cameras from the ground. The results were satisfactory, and they laid to rest the concern about instability in the transition region. It is important that the transition itself is of very short duration, since the actual vehicles are under strong acceleration. (Note, that all the curves in Figure 9 have been drawn solidly through M = 1, in order to have consistent values for aerodynamic, structural and trajectory calculations. They present the best estimate of the aerodynamics and shall not imply that these values have been obtained.)

Figure 7 A-3 in a test stand.

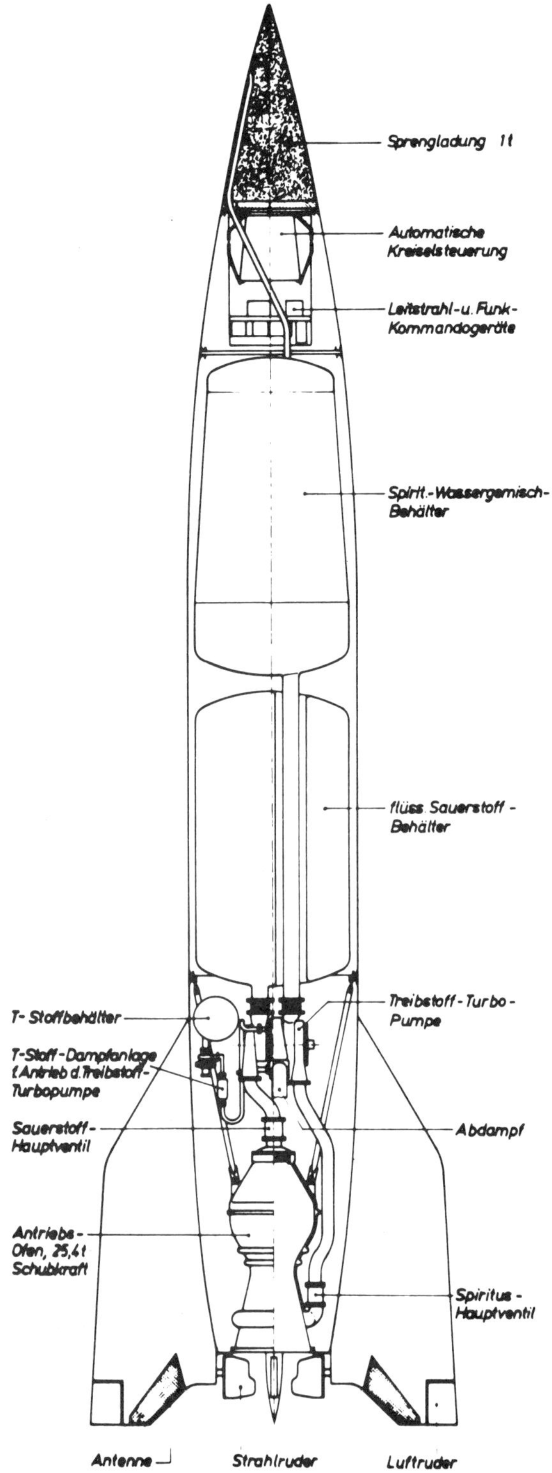

Figure 8 Configuration of German A-4.

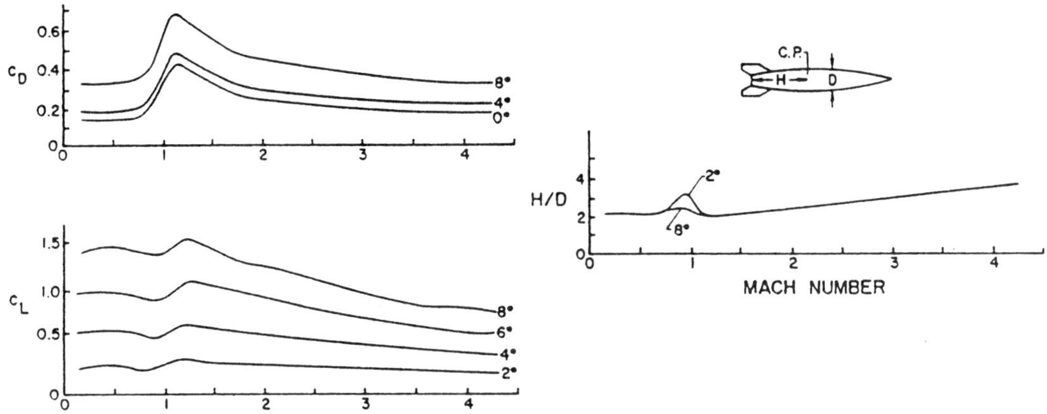

Figure 9 Drag, lift, and center pressure for A-4

Stability and Control. For aerodynamic stability, the difference Δ = C.G. – C.P. (measured from base of the vehicle) must be positive. However, it should be as small as possible, otherwise the vehicle cannot be controlled easily, since the control moment is proportional to Δ. When Δ is small, the air rudders can be small and then the hinge moment to turn the rudders were small—that is, the power was small, and weight was saved in construction of rudders, motors, and energy storage (batteries). The curves C.P. as function of M (or of time travelled) and C.G. as function of time travelled must be established very exactly, in order to fulfill the requirement for a small Δ. For A-4, Δ was mostly on the order of 0.3 of the diameter (0.3D); however, at M = 5.2 (end of powered trajectory) Δ was suspected to be 0.2 to 0.1D.

Note, that the C.P. position can be reliably measured with the 3-component balance only to about α = ± 2°. Towards α = 0° the value becomes indefinite, 0/0. Due to the importance of stability, several other independent methods had been developed but cannot be listed here.

Pressure Distribution Over A-4. Aerodynamic pressure distribution must be measured over the body, fins and air rudders, to obtain local loading needed for the structural design. One hundred-twenty finely drilled holes were used on the model, connected by 120 small tubes to 120 manometers. In the supersonic range a "half-model" method was developed. The model was bisected along its major axis and was backed by a flat plate with sharp leading edges. This enabled the attachment of the large number of tubes to the relatively small models used for supersonic testing. Nine velocities, sixteen angles of attack (α = ± 8°) were covered, each point measured four or five times, resulting in about 120,000 data points, all processed with great care by hand, since printouts or automatic calculators were unavailable. By integrating the measured pressure distribution along the body, the normal force distribution (per unit length of axis) for each of the above parameters was obtained, fulfilling the demands of the design engineers.

Rocket-Powered Vehicles Group (2)

A photograph of the A-4B is shown in Figure 10. Note the A-4B's considerable wing area, to produce lift for certain missions, such as long-range gliding or anti-aircraft maneuverability. It was stabilized either by fin assembly at the very tail, or stabilization was taken over by the rear portion of the wing, when a delta wing was used, as in a V13e.

Figure 10 German A-4B.

"Glider A-9"-Type Vehicles

These rockets had different names at different times for different selling purposes, such as: "Flossengeschoss" (fin projectile) A-4, glider A-4, A-4b, A-9. Development started in 1940. Requirements were stringent: take the body of A-4, keep the total weight of propellants and of warhead the same as A-4 and attach wings. First wind tunnel tests were carried out in the summer of 1940 on type A-4 V12a, with trapezoidal wings without sweep and on A-4 V12c, with 45° swept wings. As an example, V12a had a lift/drag ratio of 4.2 at M = 1.86. The V12c was 10% better in L/D. Figure 11 shows V12c with two trajectories. Trajectory I has a propellant cut-off angle similar to A-4, resulting in a 80 km ceiling, oscillating trajectory and little gain in range. Trajectory II has a smaller angle, resulting in a 40 km ceiling with proper gliding trajectory, resulting in maximum range of 450-500 km. V12c had a good L/D ratio, but a motion of C.P. between subsonic and supersonic range of more than one diameter, almost prohibitive. Also, at certain angles of attack, the stability and controllability with the rudders were weak, since the fins were in a downwash of the wings. Hence, other glider-types were investigated. A-4 V13e was a delta wing, with curved leading edges, showing L/D ratio 20% smaller than for V12c. A delta wing, broken into three steps, A-4 V14f ("Stufen-Gleiter") produced lift on every step, with the third step (behind the C.G.) producing a stabilizing aerodynamic moment. The L/D ratio was a little less than for V12c but the travel of C.P. with Mach number was much less—an important characteristic.

In order to develop aerodynamic rudders with maximum control effect (maximum lift on the rudder), with minimum hinge moments, pressure distribution measurements along the rudder were conducted with great success, both for the A-9 glider and for "Wasserfall."

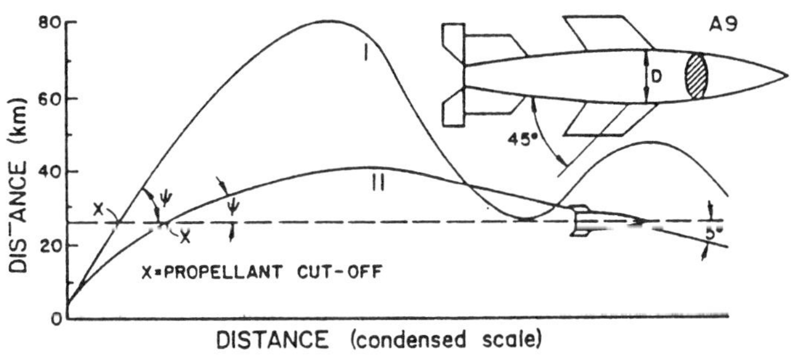

Figure 11 "Glider A-9" with two calculated trajectories.

"Wasserfall"—A Short Summary

The "Wasserfall" belonged to Group (2), as a rocket-powered vehicle, with considerable wing area and stabilized by an extra fin assembly; see Figure 3 for a wind tunnel model. The "Wasserfall" flew aerodynamically stable with subsonic, transonic speeds, developed up to M = 3. As an anti-aircraft missile, and in order to maneuver in sharp curves in all directions and from all positions, four wings were necessary, and large air rudders were placed at the tail. In the subsonic region, about 15° angle of attack was necessary; in supersonic region, about 8°. Many models with various wings, fin assemblies and rudder forms were tested. The first type used the wing cross twisted 45° against the fin assembly cross to avoid that the fins and rudders would be in the trailing vortex street of the wings. This is correct for a small angle of attack, but not for the large ones needed. One type used a concentric ring as a wing, being aerodynamically identical for any angle of attack. The final type was the wing cross and fin cross parallel to each other. The greatest achievement was that the Δ = C.G. – C.P. for the total Mach number range from 0 to 3.0 was almost constant, about Δ = 0.2D for angle of attack from 2° to 8°.

Mission requirements in trajectory and flight mechanics, aerodynamic qualities to fulfill those requirements, rudder effectiveness to produce the moments, hinge moments aerodynamically needed, but limited by the servosystem and power available—all relations are so complex, that only a detailed treatment can do justice. Here, this summary must suffice. Important is the fact that after 16 months of intensive work at the wind tunnels first in Peenemünde, then in Kochel, all requirements were fulfilled. (First report was written in April 1943; major work was completed in August of 1944; some additional details on rudders through December 1944.)

Influence of the Jet Exhaust on the Aerodynamics of the Vehicles

The influence of the jet exhaust on the aerodynamics was significant, but different in the subsonic and supersonic arenas. It requires theoretical understanding of the jet expansion from the pressure simulating the rocket chamber combustion pressure to the static pressure of the surrounding air at a given flight Mach number. Those expansions are theoretically known for the two-dimensional case, and expansion into air at rest. Here, the case is two-dimensional only at the corner, then it is axisymmetric. The expanding jet flow is hit by the outside flow, producing shock waves as on solid cones. An equilibrium is established between the expanding cone flow and the pressure behind the shock wave. Finding out the correct simulation parameters between free flight and wind tunnel test is complex. The experimental difficulties of simulation are so large that only "cold" combustion chamber air was used, not hot combustion gases. Even then, the challenge of measurement of drag and stability was carried out. The mounting of the models was from the top, near the center of gravity, using a hollow pipe, for feeding the simulated exhaust jet. The model was freely movable around the pipe as vertical axis. This arrangement was basically the same for subsonic and supersonic testing, except for the large size of the tunnel and model in subsonic testing and the small size in transonic and supersonic testing (Figure 12).

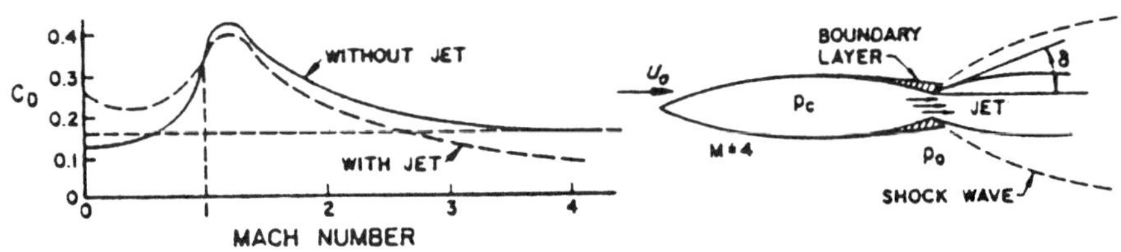

Figure 12 Drag coefficient of A-4 and jet interaction.

Effect on Drag

The subsonic region was covered from M = 0.13 through M = 0.85. At M = 0.13, the A-5 was used (which is nearly geometrically identical to A-4). The exhaust jet had M = 1.5. Pressure distribution with small holes, attached to manometers, along six meridians was obtained. Without exhaust jet there is over-pressure on the nose, some negative pressure along the cylindrical portion and slight positive pressure at the converging tail portion—the first and third contribute to the drag. With exhaust jet the surrounding air is induced like a jet pump, resulting in strong negative pressure at the tail. This results in a drag increase at subsonic speeds, see Figure 12, which is $\Delta C_D = +0.13$, nearly 100% at M = 0.13, but nearly disappears at transonic speeds. At supersonic speeds the effect is opposite, and the drag decrease about $\Delta C_D = -0.05$ for higher Mach numbers. Note that in this region the drag is decreased nearly 50 percent. The reason is an interaction of the shock wave, produced by the expanding exhaust jet, with the boundary layer at the convergent tail portion. This results in a pressure increase on that portion, equivalent to a decrease of C_D.

Effect on Stability

The evaluation of the pressure distribution mentioned above showed that the C.P. was displaced about 0.5D towards the tail with the exhaust jet on, causing the A-5 (or A-4) to become more stable. This is proven by the experiment, permitting free oscillations of the model around its C.G., after elongation to a certain initial angle, here 15°, see Figure 13. The faster oscillation means larger aerodynamic moments, also larger stability. The larger model velocities are causing faster damping which is also very welcome. Oscillation experiments were also carried out in the supersonic range for the A-4, the glider A-4 V12c, and the "Wasserfall."

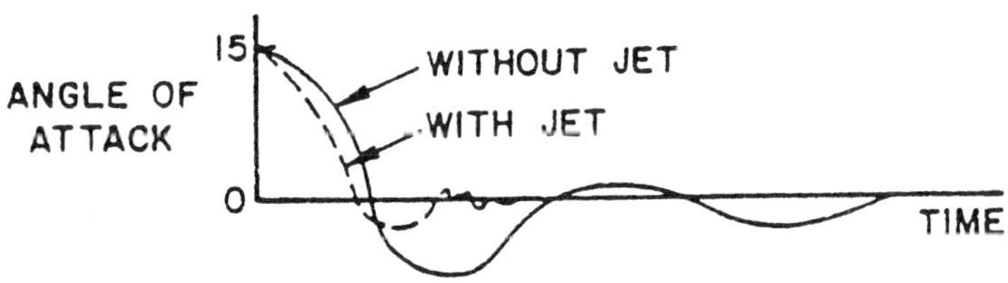

Figure 13 Stability experiments on A-5.

Conclusion

Two distinguished persons were mentioned in the beginning. I avoided mentioning others since the list would be very long. My deepest gratitude however goes to Walter Dornberger. From 1937 through 1945, in all his positions as Chief of a Department of the Army Ordnance, later as commander at Peenemünde, and later as Chief of all rocket development, he was always my direct superior. He was a profound engineer and a man of technical vision, energy, and leadership. This work could never have been accomplished without his constant personal interest, his guidance, sharp criticism, encouragement and constant support. He continued as a friend over the years.

Published reports by the author are available from 1936 through 1945, with several of summary nature. About 200 reports by members of the Wind Tunnel Group 66 are recorded as Peenemünde or Kochel Archive Reports numbered 66/1 (January 1938) through 66/205 (September 1945). Copies of all are in the possession of the author.

Part IV

Rocketry and Astronautics After 1945

Chapter 6

The Origins of Magnetospheric Physics[1]

James A. Van Allen[2]

As of 1981, magnetospheric physics is a massive and flourishing science which engages the efforts of over a thousand investigators in at least twenty different countries. The current rate of publication in this field is on the order of two or three original research papers per day. The scientific heritage of the subject lies principally in the areas of geomagnetism, aurorae, geomagnetic storms, and the geophysical aspects of cosmic rays and solar corpuscular streams. This heritage is well represented by a wide body of important scientific literature.[3]

The general magnetic field of the Earth is essential to the existence of its magnetosphere. Indeed the magnetosphere of the Earth or of any other celestial body may be defined as that region surrounding the body within which its magnetic field, however distorted by external currents, controls the motion of electrically charged particles. The magnetosphere of the Earth encompasses a huge population of such particles—electrons, protons, and other ions—whose source and gross dynamics are traceable principally, but

---

[1] Presented at the Fifteenth History Symposium of the International Academy of Astronautics, Rome, Italy, 1981.

[2] Department of Physics, University of Iowa, Iowa City, Iowa, U.S.A. I am very grateful for a Regents' Fellowship of the Smithsonian Institution for the period January-August 1981. During this time I was in residence at the National Air and Space Museum in Washington, D.C., writing a monograph entitled "Origins of Magnetospheric Physics." The present paper is an abridgement of a portion of that monograph.

[3] S. Chapman, and J. Bartels, *Geomagnetism* (Oxford, England: The Clarendon Press, 1940, 2 vols.); L. Jánossy, *Cosmic Rays* (Oxford, England: The Clarendon Press, 1948); S. K. Mitra, *The Upper Atmosphere* (Calcutta, India: The Asiatic Society, 1952 ed.); D. J. X. Montgomery, *Cosmic Ray Physics* (Princeton, NJ: Princeton University Press, 1949); C. Störmer, *The Polar Aurora* (Oxford, England: The Clarendon Press, 1955); H. Alfvén, *Cosmical Electrodynamics* (Oxford, England: The Clarendon Press, 1950).

not wholly, to the solar wind, a magnetized, ionized gas emitted by the Sun and flowing outward through the solar system.

On the one hand, magnetospheric physics is closely related to laboratory plasma physics, and, on the other, it is presumably applicable to natural phenomena in planetary and astrophysical systems throughout the universe. The theory of ionized gases in magnetic and electric fields provides the unifying principles of the subject over a vast range of physical scale. The present sophistication of magnetospheric physics stands in stark contrast to its primitive beginnings.

Precursory Work

During the period 1946-1957, a large number of investigations were conducted in the upper atmosphere of the Earth with scientific instruments on rockets flown more or less vertically to altitudes of up to 390 kilometers by research workers in the United States, the Soviet Union, the United Kingdom, Australia, France, and Japan.[4] Some of these flights were devoted to the study of the intensity and nature of the cosmic radiation above the appreciable atmosphere. Such measurements were extended over the latitude range 77° North to 71° South by an inexpensive balloon-launched rocket technique during the latter part of this period.[5] This series of flights from ships at sea also resulted in the first direct observations of the primary auroral radiation.[6]

Experience in conducting scientific measurements with rocket-borne instruments, *and* the rapid development of high performance rockets following World War II, made it realistic to plan scientific work with Earth orbiting satellites as part of the 1957-1958 International Geophysical Year. The first flight of an artificial satellite of the Earth, *Sputnik I*, was achieved by the Soviet Union on 4 October 1957. The radio signals from *Sputnik I*, at 20.005 and 40.002 MHz, were received by stations throughout the world. The distinctive—beep-beep-beep—modulation of these signals heralded the beginning of a new scientific epoch. *Sputnik II*, launched on 3 November 1957, carried the first detectors of energetic charged particles to be flown on a satellite. Two shielded (10 g cm^{-2}) Geiger tubes (gas-discharged counters 10 cm in length and 1.8 cm in diameter) in this payload were intended to provide a geographically comprehensive survey of the cosmic ray intensity above the atmosphere. This flight obtained observations over the Soviet Union for a period of about seven days in the altitude range 225 to 700 km, the latitude

[4] Homer E. Newell, Jr., *High Altitude Rocket Research* (New York: Academic Press, 1953).

[5] James A. Van Allen, "Balloon-Launched Rockets for High-Altitude Research," in Homer E. Newell, Jr., ed., *Sounding Rockets* (New York: McGraw-Hill Book Co., 1959); James A. Van Allen and M. B. Gottlieb, "The Inexpensive Attainment of High Altitudes with Balloon-Launched Rockets," in R. L. F. Boyd and M. V. Seaton, eds., *Rocket Exploration of the Upper Atmosphere* (London, England: Pergamon Press, 1954).

[6] L. H. Meredith, M. B. Gottlieb, and James A. Van Allen, "Direct Detection of Soft Radiation Above 50 Kilometers in the Auroral Zone," *Physics Review*, 97 (1955): pp. 201-205; James A. Van Allen, "Direct Detection of Auroral Radiation with Rocket Equipment," *Proceedings of the National Academy of Sciences*, 43 (1957): pp. 57-62.

range 40° to 65° North and the longitude range 25° to 143° East. Reliable data were obtained on the latitude and altitude dependence of cosmic ray intensity. On a particular pass on 7 November 1957 a brief series of "bursts" of substantially greater (~50 percent) than average intensity was observed coherently by the two detectors at latitudes greater than 58° North (Figure 1). This effect was termed a "cosmic-ray burst," but no physical interpretation was suggested. Otherwise the *Sputnik II* data exhibited a smooth dependence of previous rocket measurements at lower altitudes.[7]

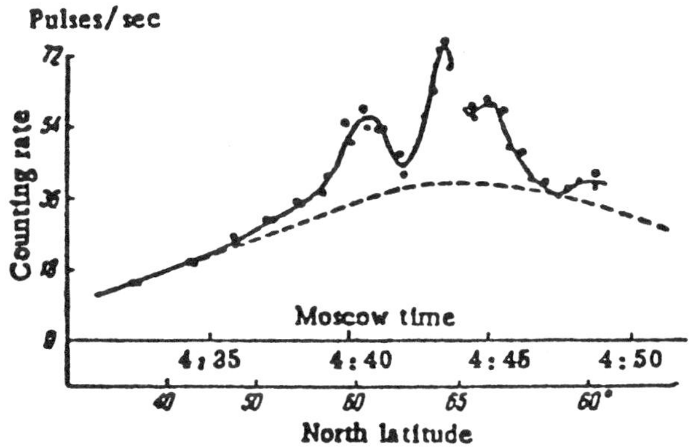

Figure 1 Time dependence of the counting rate of a Geiger tube on *Sputnik II*. The dashed curve shows the usual case and the points and solid curve show the occurrence of an unusual "cosmic ray burst" on 7 November 1957 (See S. N. Vernov, N. L. Grigorov, Iu. I. Logachev, and A. E. Chudakov, "Cosmic Radiation Measured on the Second Artificial Satellite," *Doklady Academii Nauk, U.S.S.R.*, 120 (1958): 1231-1233, also published in English, *Soviet Physics Doklady*, 3 (1958): 617-619.

Discovery of the Inner Radiation Belt of the Earth

In anticipation of the opportunity for extending our sparse geographical survey with rockets of the cosmic ray intensity above the atmosphere, I had proposed the flight of a simple radiation detector on an early U.S. Earth satellite.[8] The instrumentation prepared for this purpose at the University of Iowa had a single, lightly-shielded (1.5 g cm^{-2}) Geiger tube (10.2 cm length and 2.0 cm diameter) as the basic detector. One such instrument was a part of the payload (Figure 2) of the first U. S. satellite, *Explorer I*, launched on 1 February 1958 (UT), 31 January, EST from Cape Canaveral, Florida, into an orbit inclined at 33° to the equator with perigee and apogee altitudes of 360 and 2550 km, respectively. A similar instrument was flown in *Explorer III*, launched on 26 March

[7] S. N. Vernov, N. L. Grigorov, Iu. I. Logachev, and A. E. Chudakov, "Cosmic Radiation Measured on the Second Artificial Satellite," *Doklady Academii Nauk, USSR*, 120 (1958): pp. 1231-1233, also published in English, *Soviet Physics Doklady*, 3 (1958): pp. 617-619.

[8] James A. Van Allen, "Cosmic-Ray Observations in Earth Satellites," in James A. Van Allen, ed., *Scientific Uses of Earth Satellites* (Ann Arbor: University of Michigan Press, 1956), pp. 171-187.

1958 into an orbit of the same inclination with perigee and apogee altitudes of 188 and 2800 km, respectively.[9] The *Explorer III* instrument included the important addition of a miniature magnetic tape recorder which stored the counting rate data for 135 minutes (more than a complete orbital period) and then, on radio command, played back the stored data over the telemetry transmitter in six seconds while within the range of a single receiving station.[10]

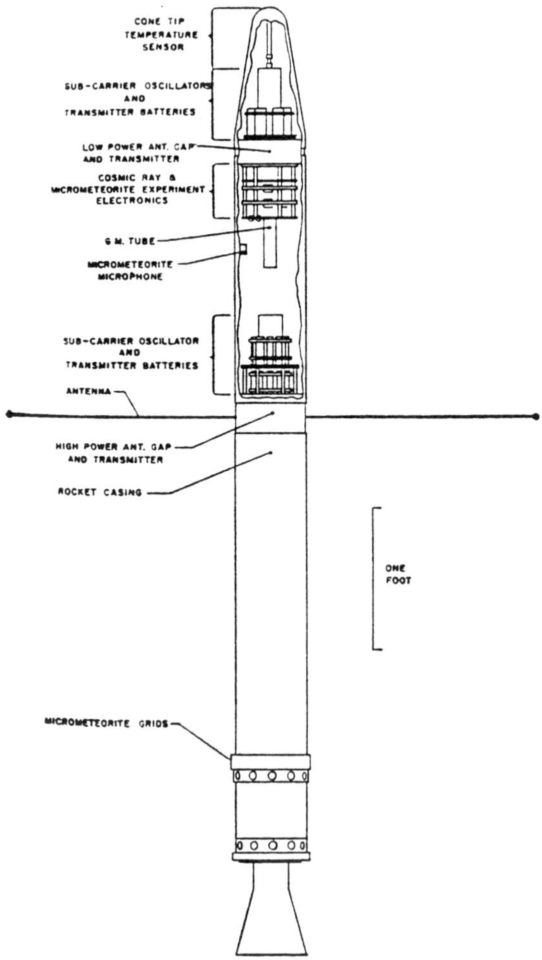

Figure 2 Outline drawing of the payload of *Explorer I* (1958 Alpha), as permanently attached to the fourth stage rocket of the propulsion system (See George H. Ludwig, "Cosmic-Ray Instrumentation in the First U.S. Earth Satellite," *Review of Scientific Instrumentation*, 30 (1959): 223-29).

[9] George H. Ludwig, "Cosmic-Ray Instrumentation in the First U.S. Earth Satellite," *Review of Scientific Instrumentation*, 30 (1959): pp. 223-229.

[10] George H. Ludwig, "The Instrumentation in Earth Satellite 1958 Gamma," M. S. Thesis, University of Iowa Research Report 59-3, 1959. See also *Satellites 1958 Alpha and Gamma: High Intensity Radiation Research and Instrumentation*, IGY Satellite Report No. 13 (Washington, DC: National Academy of Sciences, 1961), pp. 31-93.

Within the first few weeks of *Explorer I*'s orbital flight, we had only a rather sparse set of data, consisting of segments of the order of one minute duration from many different and somewhat uncertain positions in latitude, longitude, and altitude. During some segments of data, the counting rate of our Geiger tube was of the order of 20 to 100 counts per second, generally within the range that we had expected for cosmic radiation, and the instrument appeared to be operating reliably. In other segments of data there were no counts observed for as long as two minutes. On one noteworthy pass the rate underwent a transition from zero to a reasonable value within about twenty seconds. There was no conceivable way in which the cosmic ray intensity could drop to zero at high altitudes. On the other hand, we had a high level of confidence in the Geiger tube and the associated tuning fork timer and electronic circuitry by virtue of conservative design and the rigorous thermal and mechanical conditions that it had survived in the pre-flight testing program. This puzzle hung over our heads as we tried to find if this strange effect had any systematic dependence on passing through the Earth's shadow, on payload temperature, or on altitude, latitude or longitude. Noise-free data accumulated slowly.

Throughout the few weeks following the launch of *Explorer I* we were heavily occupied in preparing *Explorer II* (lost in the launch failure of 5 March) and *Explorer III*; in developing data reduction and analysis techniques, to which we had given relatively little attention before flight; in formulating plans for subsequent flights; and in coping with a steady flow of telephone calls on practical arrangements and on inquiries concerning our progress. Our original plan was to accumulate a comprehensive body of data on the distribution of cosmic-ray intensity around the Earth on a leisurely basis. But the widespread interest in *Explorer I* produced an urgent demand for an immediate report of observational results.

Soon after the launching of *Explorer III*, I flew to Washington, D.C., to confer with Joseph Siry, John Mengel, and others at the Naval Research Laboratory (NRL) and pick up preliminary orbital data for *Explorer III*. Contrary to the implications of some popular accounts, the Vanguard group fully supported the Explorer program in many vital ways. The first successful launch of a Vanguard satellite had occurred on 17 March 1958, and the NRL team was operating on an around-the-clock basis, now handling the tracking and data acquisition for three satellites. From NRL, I returned to the Vanguard data center on Pennsylvania Avenue and picked up the complete record of a successful playback of data from our *Explorer III* tape recorder. The playback had been received at the San Diego minitrack station on 28 March 1958.

I put the record in my briefcase and returned to my room in the Dupont Plaza Hotel. There, with the aid of graph paper, my slide rule, and a ruler from a nearby Peoples Drug Store, I worked out the counting rates of our Geiger tube as a function of time for a full 102-minute period and plotted them (Figure 3). The data provided a beautiful confirmation and extension of the fragmentary data from *Explorer I*. The counting rate at low altitudes was in the expected range of 15 to 20 counts per second. There was then a very rapid increase to a rate exceeding 128 counts per second (the maximum recordable rate of our on-board storage system). A few minutes later the rate decreased rapidly to zero. Then after about fifteen minutes it rose rapidly from zero to greater than 128 counts per second again and remained high for forty-five minutes, then again decreased rapidly to 18 counts per second as the orbit around the Earth was nearly completed. At 3:00 a.m. I packed my work sheets and graph and turned in for the night with the conviction that our instruments on both *Explorers I* and *III* were working reliably and giving reproducible results, but that we were encountering a mysterious physical effect of a real nature.

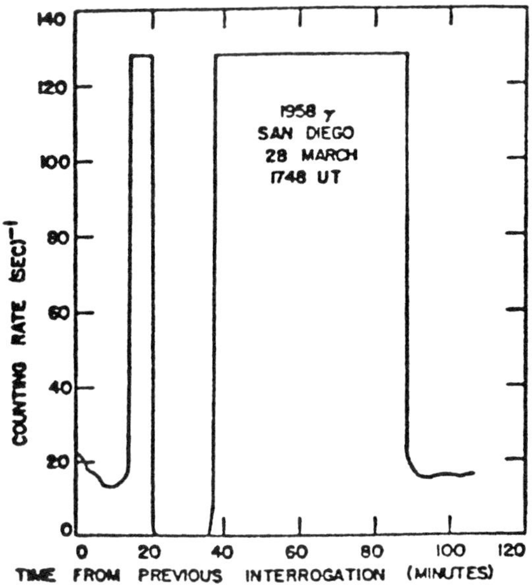

Figure 3 *Apparent* counting rate of the Geiger tube on *Explorer III* (1958 Gamma) as derived from a tape recording for one orbit around the Earth. Note that 128 count sec^{-1} is the saturation level of the recording system and that a zero *apparent* counting rate corresponds to a *true* counting rate greater than 25,000 count sec^{1} (James A. Van Allen, "Observations of High Intensity Radiation by Satellites 1958 Alpha and 1958 Gamma, *Satellites 1958 Alpha and Gamma: High Intensity Radiation Research and Instrumentation*, IGY Satellite Report No. 13 (Washington, DC: National Academy of Sciences, 1961), pp. 1-22.

Early the following day I flew back to Iowa City and proudly displayed my graph and my conviction to my colleagues, Earnest Ray and Carl McIlwain. Meanwhile, it had occurred to McIlwain that a sufficiently great intensity of radiation would overload our detection system and drive the "apparent" counting rate to zero, a fact that we all knew but had temporarily ignored, because of our predisposition to reject the possibility of such great intensities, and because of our apprehension about possible malfunctioning of the instrument. During the previous day, McIlwain made tests with our prototype Geiger tube system using a small x-ray machine and demonstrated that a true rate exceeding about 25,000 counts per second would indeed result in an apparent telemetered rate of zero. The conclusion was immediate: At altitudes of the order of 1200 km and greater the radiation intensity was at least a thousand times as great as that of cosmic radiation!

Ray's famous, though consciously inaccurate remark summarized the situation: "My God, space is radioactive!" Our realization that there was a very great intensity of radiation at high altitudes rationalized our entire body of data. George Ludwig returned from the Jet Propulsion Laboratory to the University of Iowa on the 11th of April 1958, and the four of us continued working feverishly in analyzing the data from *Explorers I* and *III* (using primitive hand methods) and organizing them by altitude, latitude, and longitude. The systematic, repetitive run of the data on successive orbits confirmed my earlier conclusion that both instruments were indeed operating reliably.

During mid-April we prepared graphs and a short written statement of our raw findings and I mulled over their meaning.[11] I entertained two quite different lines of thought: (a) that we might be detecting x rays or gamma rays, possibly from the Sun or (b) that the radiation at high altitudes might be akin to the auroral soft radiation which we had studied during the preceding several years with rockroom flights at high latitudes and most recently by McIlwain's rocket flights from Fort Churchill. We had identified the auroral radiations as being principally electrons having energies of the order of 10s of keV. I quickly rejected hypothesis (a) on the conclusive grounds that the high intensity radiation was present during both daylight and dark conditions, that it exhibited a latitude dependence, and that the extremely rapid increase in intensity with increasing altitude was impossible for any type of electromagnetic radiation. Specifically, the rate of increase of intensity was a factor of ten within an altitude range of less than 100 kilometers at an altitude of the order of 100km; the decrease in atmospheric "thickness" within that increment of altitude was totally negligible compared to the same 1.5 g cm^{-2} of material in the nose cone and wall of the detector. Such an increase could not be due to decreased absorption of x-rays or gamma rays of any wavelength. Hence, I concluded that the effect must be attributed to electrically charged particles, constrained by the Earth's external magnetic field from reaching lower altitudes.

Such particles could not be arriving directly from infinity because of the concentration of the effect near the geomagnetic equator, in complete defiance of C. Störmer's theory of the entrance of electrically charged particles from a distant source into the Earth's magnetic field. By virtue of my familiarity with a 1907 paper by Störmer on the bound motion of an electrically charged particle in a dipolar magnetic field, with magnetic field confinement of charged particles in laboratory machines during my 1953-1954 work building and operating an early version of a stellarator at Princeton University, and with an early 1958 suggestion of N. C. Christofilos (discussed later), I further concluded that the causative particles were present in trapped orbits in the geomagnetic field, moving in spiral paths back and forth between the northern and southern hemispheres and drifting slowly around the Earth.[12] The intensity of these trapped particles, whose mirror points were at low altitudes, would be diminished by atmospheric absorption. I called this population of particles geomagnetically trapped corpuscular radiation. Later it was termed the "radiation belt" of the Earth because of the toroidal form of the populated region, encircling the Earth like a belt.

---

[11] James A. Van Allen, George H. Ludwig, Ernest C. Ray, and Carl E. McIlwain, "Observation of High Intensity Radiation of Satellites 1958 Alpha and Gamma," *Jet Propulsion*, September 1958, pp. 588-592; and "Radiation Observation with Satellite 1958 e," *Journal of Geophysical Research*, 64 (1958): pp. 271-286.

[12] C. Störmer, "Sur les trajectoires des corpuscles electrises dans l'espace sous l'action du magnetisme terrestre," Series IV, *Arch. Sci. Phys. Nat.*, 24 (1907): pp. 317-364.

The foregoing account of observations and interpretation is essentially the one that I gave in a special joint session of the American Physical Society and the National Academy of Sciences in the latter's auditorium on 1 May 1958. A large press conference was held following the lectures with me and other participants in the *Explorer I* and *Explorer III* program, and the results were reported extensively by news media on the following day. Fortunately, and unknown to me until more than a year later, a "Voice of America" tape recording was made of my lecture and of the ensuing question and answer period. A written transcription of this tape, together with copies of my illustrations, provided a documented, published record of this lecture, complete with grammatical errors and colloquial language.[13]

I had adopted at that time the working hypothesis that the trapped radiation consisted of "electrons and likely protons, energies of the order of 100 keV and down, mean energies probably about 30 keV." In this vein of thought, the response of our Geiger tube would be attributed to *bremsstrahlung* produced as the electrons bombarded the nose cone of the instrument. If this *bremsstrahlung* interpretation was correct, I estimated that an omnidirectional intensity of 10^8 to 10^9 $(cm^2\ sec)^{-1}$ of 40 keV electrons would be required to account for the counting rates at altitudes of ~ 1500 km over the equator. However, in my 1 May lecture, as well as in response to a question at the end, I emphasized that we had no definitive identification of particle species, and that the particles might be penetrating protons or electrons. I did, however, regard protons and electrons of energies necessary to penetrate the Geiger tube directly, namely $E_p > 35$ MeV and $E_e > 3$ MeV, as unlikely in view of our auroral zone measurements with rocket-borne equipment. Later in 1958 we found that this opinion was incorrect.

Of the some 1,500 real-time recordings of *Explorer I* telemetry signals in the period 1 February to 9 May 1958, 850 contained readable data. A much larger body of data was obtained from *Explorer III*. During its 44 days of useful life, the satellite completed 523 orbits around the Earth and 504 tape recorder play-backs were attempted. Of these attempts a total of 408, or 81 percent, were successful. The tape recorder operated perfectly throughout this period and never failed to respond to a command that was electronically successful.

Virtually all of the *Explorer I* data were reduced and analyzed (Figures 4 and 5).[14] But only some 10 percent of the *Explorer III* data were reduced. Even as of 1981, George Ludwig and I regret not having completed the reduction of the major part of both the real-time and stored data from *Explorer III*. In early 1958, however, we considered that we had gleaned the principal elements of the data and hardly felt it worthwhile to continue because of the limited dynamic range of the instrument and our preoccupation with follow-on satellite missions carrying detectors specifically designed for the then-known intensities and for particle species identification and other more revealing properties of the radiation.

---

[13] James A. Van Allen, "Observations of High Intensity Radiation by Satellites 1958 Alpha and 1958 Gamma," in *Satellites 1958 Alpha and Gamma: High Intensity Radiation Research and Instrumentation*, IGY Satellite Report No. 13 (Washington, DC: National Academy of Sciences, 1961), pp. 1-22.

[14] S. Yoshida, George H. Ludwig, and James A. Van Allen, "Distribution of Trapped Radiation in the Geomagnetic Field," *Journal of Geophysical Research*, 65 (1960): pp. 807-813.

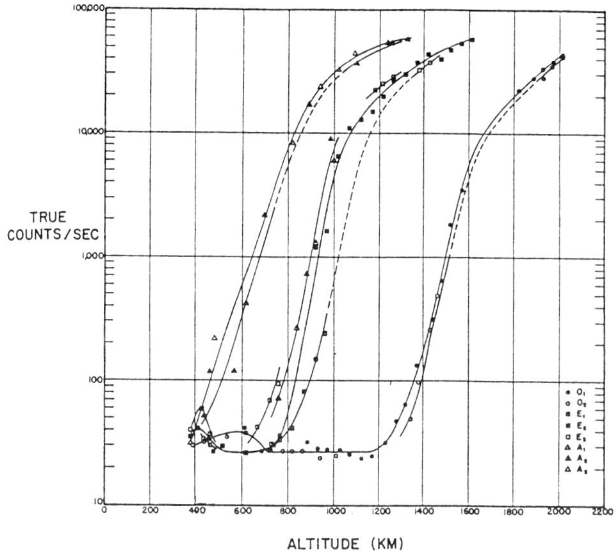

Figure 4 Altitude dependence of the *true* counting rate of the Geiger tube on *Explorer I* as observed within 19° of the magnetic dip equator at different geographic longitudes approximately as follows: two curves on the left, 288° East; three curves in the center, 5° East; and two curves on the right 105° East (See S. Yoshida, George H. Ludwig, and James A. Van Allen "Distribution of Trapped Radiation in the Geomagnetic Field," *Journal of Geophysical Research*, 65 (1960): 807-813).

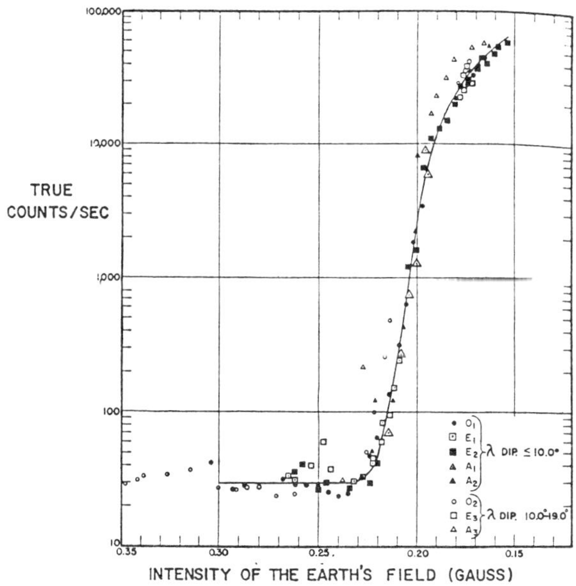

Figure 5 All of the counting rate data of Figure 4 replotted as a function of scalar magnetic field strength, illustrating the validity of the geomagnetic trapping hypothesis. (See S. Yoshida, George H. Ludwig, and James A. Van Allen "Distribution of Trapped Radiation in the Geomagnetic Field," *Journal of Geophysical Research*, 65 (1960): 807-813).

There is no evidence, so far as I can find, that Soviet scientists were aware of the existence of geomagnetically trapped radiation before my 1 May 1958 report of *Explorer I/III* observations. It may be noted that the high intensity radiation found by these two satellites was a part of what was later recognized as the *inner* of two major radiation belts.

Radiation Observations with Sputnik III

The third Soviet satellite, *Sputnik III*, was launched on 15 May 1958 into an orbit inclined at 65° to the equator. Its orbit, similar to that of *Sputnik II*, had perigee and apogee altitudes of 225 and 1,880 km, respectively. Among other equipment, *Sputnik III* carried a scintillation detector for the observation of photons and charged particles. The scintillator was a cylinder of crystalline NaI (T*l*) 4.0 by 3.9 cm in size, mounted on a photomultiplier tube whose photocathode had a diameter of 4.0 cm. The effective shielding of the detector was at least 1 g cm^{-2}. Three quantities were transmitted: (a) the counting rate of pulses corresponding to energy release in the crystal of 35 keV or more, (b) the anode current of the photomultiplier tube, and (c) the current to an intermediate dynode. A preliminary report of the observations with this instrument was made in a special lecture by S. N. Vernov at the Fifth General Assembly of CSAGI (Special Committee for the International Geophysical Year) in Moscow, 30 July-9 August 1958. Among U.S. delegates to this meeting were Ernest Ray of the Iowa group, who gave a preliminary report of *Explorer IV* results obtained during the first few days of its flight (see later section). The first written report (available in English) of *Sputnik III* observations that I have found was submitted to *Planetary and Space Science* on 17 January 1959 and published soon thereafter.[15] The stated intention of this investigation was "to obtain data on photons at high altitudes." The choice of detector and associated data handling system does not suggest any prior knowledge of the existence of geomagnetically trapped radiation, nor is there any hint of such prior knowledge in the text of the paper. Nonetheless, as with *Explorer I*, the principal results related to electrically charged particles (in part, via their *bremsstrahlung*).

Vernov and his colleagues distinguished two observed zones of high intensity radiation—a polar zone (latitudes 52° to 65° North) and an equatorial zone. In the section of their paper entitled "Electronic Component of Cosmic Rays in the Polar Region," the authors stated, in part:

> Comparing this increase of ionization and increase of the counting rate, one can estimate photon energy of the bremsstrahlung radiation and, consequently, the energy of electrons. The photon energy will be of the order of 100 keV or less. The sign "less" arises due to the fact that in some cases only a low limit of counting rate is known. However, the photon energy cannot be significantly lower than 100 keV, since the threshold is 35 keV. Thus, the most probable value of the energy of elec-

[15] The following quotes and my observations on the findings are from S. N. Vernov, A. E. Chudakov, E. V. Gorchakov, J. L. Logachev, and P. V. Vakulov, "Study of the Cosmic-Ray Soft Component by the 4th Soviet Earth Satellite," *Planetary and Space Sciences*, 1 (1959): pp. 86-93.

trons responsible for the effect is about 100 keV. . . . Thus, the polar region is characterized by a permanent electron flux, the intensity of which varies in a wide range. Taking into account the efficiency of electron registration by the bremsstrahlung radiation, we estimate the electron flux as 10^3 to 10^4 particles (cm^2 sec sterad)$^{-1}$. . . .

The authors concluded this section of their paper as follows:

> At present it is difficult to give a complete interpretation of the electron component observed. It is not excluded that these electrons are accelerated near the Earth by electric fields similar to those assumed to be in the aurora. But it is possible also that the electron component comes from far away, for example, from the Sun; and penetrates, despite the small amount of energy of the particles, through the Earth's magnetic field due to the difference between the field and that of an ideal dipole.

On a purely observational basis the *Sputnik III* data represented discovery of the Earth's *outer* radiation belt, inasmuch as they were acquired before those of *Explorer IV* and *Pioneer III* (see later section). However, the interpretation of Vernov and his associates did not encompass the idea of geomagnetic trapping for what they called the "electron component of the cosmic rays in the polar region."

The next section of this important paper was entitled "Zone of High Intensity in the Equator Region." Therein the authors confirmed our *Explorers I/III* results and added significant detail on latitude and altitude dependence of intensity. They also found the radiation intensity in the equatorial region to have approximate north-south symmetry relative to the geomagnetic equator, and to be stable in time. The time stability at low latitudes was contrasted to the marked temporal variability at high latitudes. They estimated that the energy deposition rate in the NaI crystal was of the order of 5 erg sec^{-1} at an altitude of 1200 km near the geomagnetic equator. The authors remarked:

> Independently of the mechanism of production of particles in the equatorial zone, it is obvious that the main role in this effect is played by the factor of storing. This fact is proved by the concentration of particles in the equatorial zone where, at a sufficient altitude, they can oscillate for a very long period of time.

At the end of this section of the paper, the authors concluded as follows:

> It is possible now to make a hypothesis that the equatorial zone of high intensity is placed symmetrically to the Earth's magnetic field and is characterized by a strong concentration of the density of particle flux in the plane of magnetic equator. Such an equatorial zone provides, apparently, ideal conditions for particle oscillations in a magnetic field, and leakage is determined probably only by ionization losses and losses for radiation. In this case such a long lifetime of particles is possible (of the order of a year) that even such a weak mechanism of injection as the decay of neutrons leaving the atmosphere under the action of cosmic rays appears to be sufficient to explain the observed intensity.

The conclusion clearly embraced the concept of geomagnetic trapping as appropriate for the equatorial region and suggests that the decay of neutrons produced in the atmosphere by cosmic rays provides a possible source. However, no observational identification of the nature of the trapped particles was achieved nor was any information on intensity vs. range obtained. Also, it was not clear whether the authors were thinking of low energy or high energy neutrons or of electrons or protons or both as decay products thereof.

On *Sputnik III* there were two other detectors relevant to the present discussion. Each detector consisted of a fluorescent film of ZnS (Ag), 5 cm in diameter and 2×10^{-3} g cm^{-2} in thickness, deposited on the face of a photomultiplier tube. Aluminum foils of thickness 8×10^{-4} g cm^{-2} and 4×10^{-4} g cm^{-2} covered the fluorescent films of the respective detectors. The primary purpose of this instrument was to detect and measure corpuscular streams, especially those which cause aurorae and contribute to heating the upper atmosphere. Again the principal results related to geomagnetically trapped particles at relatively low latitudes. V. I. Krasovskii made a preliminary report on his investigation at the Fifth General Assembly of CSAGI in Moscow in 1958, including the concept of geomagnetic trapping.[16] In a 1961 full report of Krasovskii's team's findings they concluded that:

> . . . large currents of electrons with energies about 10 keV were detected at 1900 km above the southern part of the Pacific Ocean. The intensity of these currents was very high, and in the majority of cases the apparatus gave off-scale readings, since it was not expected that such high intensities would be present . . . . For all the off-scale readings, the energy flux of the electrons under investigation was in excess of 100 erg $(cm^2 \text{ sec})^{-1}$ up to 1900 km from the Earth's surface. If electron currents of this intensity were capable of penetrating into the lower parts of the atmosphere, for example the F-layer of the ionosphere, then they could not have remained unnoticed, since they would appreciably increase the ionization of the upper atmospheres and would give rise to polar aurorae. Since such phenomena were not observed, the recorded currents were explained by an oscillatory motion of the electrons along the magnetic lines of force (July 1958, Fifth CSAGI Meeting, Moscow) . . . .
>
> The majority of the recorded electrons move near the plane perpendicular to the magnetic line of force, and the flux of electrons along the line of force in the direction toward the Earth (downward direction) is greater than in the opposite direction. Electrons moving toward Earth at small angles to the magnetic line of force were also recorded, and this suggests that particles which penetrate into the lower part of the atmosphere appear as a result of some processes occurring at distances in excess of 1900 km. It was established that the energy flux of electrons with a magnetic moment enabling penetration into the F-layer of the ionosphere can reach values of about 1 erg $(cm^2 \text{ sec})^{-1}$.[17]

[16] V. I Krasovskii, Yu. M. Kushnir, G. G. Bordovskii, G. F. Zakharov, and E. M. Sveltlitskii, "The Search for Corpuscles with the Help of the Third Artificial Satellite," in *Artificial Earth Satellites* (New York: Plenum Press, 1960); 2:75-77. This was also published in *Uspekhi Fiz. Nauk*, USSR, 64 (1958): 425.

[17] V. I. Krasovskii, I. S. Shklovskii, Yu. I. Galperin, E. M. Sveltlitskii, Yu. M. Kushnir, and G. A. Bordovskii, "Discovery of Approximately 10-kev Electrons in the Upper Atmosphere," in *Artificial Earth Satellites* (New York: Plenum Press, 1961), 6:137-155.

Follow on Investigations with Explorer IV and Pioneers I, II, and III

The next satellite to carry radiation instruments was *Explorer IV*, under the joint auspices of the U.S. National Committee for the International geophysical year and the Department of Defense. Its purpose was twofold: First, to follow up on the discoveries of *Explorers I/III* with a system of detectors of adequate dynamic range to cope with the then-known intensity of trapped radiation and of more discriminating properties than those used in *Explorers I/III*, and, second, to observe the effects of a planned series of high altitude atomic bomb tests. The bomb tests had been proposed in the Atomic Energy Commission by Nicholas Christofilos in late 1957 for the purpose of producing artificial radiation belts of electrons from the decay of radioactive fission products of the bombs.[18] The University of Iowa group was entrusted with preparing the package of radiation detectors. We settled on an array of four radiation detectors:

1. *Detector A:* A circular disc of plastic scintillator, thickness 0.178 cm, diameter 0.762 cm, cemented on the face of an end-window photomultiplier tube and covered by 0.14 cm^{-2} of aluminum (pulse counting).
2. *Detector B:* A circular disc of CsI (T*l*) scintillating crystal, thickness 0.203 cm, diameter 0.762 cm, cemented on the face of an end-window photomultiplier tube, with the crystal coated with an evaporated layer of aluminum (0.2 mg cm^{-2}) and further shielded by a nickel foil of 0.8 mg cm^{-2} thickness (quasi-logarithmic electrometer for anode current).
3. *Detector C:* A miniature Geiger tube having a omnidirectional geometric factor of 0.14 cm^2 for 1.2 g cm^{-2} shielding and its full value of 0.705 cm^2 for about 5 g cm^{-2} shielding (pulse counting).
4. *Detector D:* A nearly identical Geiger tube, enclosed in an additional lead shield of 1.6 g cm^{-2} thickness so that its omnidirectional geometric factor was 0.14 cm^2 for 2.8 g cm^{-2} shielding and 0.823 cm^2 for 6 g cm^{-2} shielding (pulse counting).

Detectors A and B were clearly directional by virtue of physical collimators and the geometric shape of the sensitive elements; Detectors C and D were sensitive omnidirectionally, but there was an unavoidable variation of the shielding in different directions. Every effort was made to provide the necessary dynamic range to cope with the intensity of the natural radiation and the estimated intensity of that to be artificially injected.

Upgraded high speeds of the four-stage launching vehicle made it possible to plan an increase in the inclination of the satellite orbit, from the 33° orbits of *Explorers I* and *III*, to 50° in order to provide improved coverage in latitude. *Explorer IV* was launched from the east coast of the United states on 26 July 1958 into an orbit of 50° inclination to the equator, with perigee and apogee altitudes of 260 and 2200 km, respectively.

All the detectors operated properly and there was soon a large amount of fresh observational data available. The findings of *Explorers I/III* were massively confirmed, and major advances were made in knowledge of the distribution of intensity and the

[18] N. C. Christofilos, "The Argus Experiment," *J. of Geophysical Research* 64 (1959): pp. 869-875.

nature of the geomagnetically trapped radiation.[19] Energy fluxes as great as ~ 100 erg $(cm^2 \text{ sec sr})^{-1}$ were measured with Detector B at high altitudes near the geomagnetic equator and excellent angular distributions were obtained. Comparison of the rates of detectors C and D showed that the radiation near the geomagnetic equator was much more penetrating than at high latitudes and was not compatible with the *bremsstrahlung* interpretation that I had suggested earlier.

In agreement with the data from *Sputnik III*, the distribution of intensity as observed near the Earth was found to be bifurcated into a low latitude region and a high latitude region separated by a "slot" of lesser intensity at a geomagnetic latitude of about 48°. The angular distributions of directional intensity and the altitude distributions of omnidirectional intensity in *both* regions were characteristic of magnetically trapped radiation. This result for the high latitude region contradicted the 1958 interpretation of Vernov and his colleagues as quoted above. Contours of constant intensity turned away from the Earth at geomagnetic latitudes exceeding about 60°, suggesting that aurorae occur at or beyond the high latitude boundary of trapping. The diversity of detectors on *Explorer IV* also gave a parametric characterization of the radiation at each point along the orbit, but we were still unable to reach unique conclusions on the relative proportion of electrons and protons because of ignorance of their respective energy spectra. In a later paper I gave a variety of alternative interpretations.[20] Meanwhile, the U.S. Department of Defense had produced bursts of two 10-megaton yield atomic bombs, called *Teak* and *Orange* on 1 August and 12 August 1958, at approximate altitudes of 75 and 45 km, respectively, above Johnston Atoll in the North Pacific. Three Argus atomic bomb bursts (about 1.4 kiloton yield) were produced over the South Atlantic on 27 August, 30 August, and 6 September 1958 at altitudes of about 200, 250, and over 480 km, respectively.

We observed with *Explorer IV* the effects of all five nuclear bursts in populating the geomagnetic field with energetic electrons. Despite the large yields of *Teak* and *Orange*, the effects on the radiation belts were small and of only a few days' lifetime because of atmospheric absorption corresponding to the relatively low altitudes of ignition. The three Argus bursts produced clear and well observed effects and gave a great impetus to understanding geomagnetic trapping.[21] About three percent of the available electrons were injected into durably trapped orbits. The apparent mean life-time of the first two artificial radiation belts was about three weeks, and about a month for the third. In all three cases a well defined and durable shell of artificially injected electrons was produced. Worldwide study of these shells provided results of basic importance in giv-

---

[19] James A. Van Allen, Carl E. McIlwain, and George H. Ludwig, "Satellite Observations of Electrons Artificially Injected into the Geomagnetic Field," *Journal of Geophysical Research*, 64 (1959): pp. 272-286.

[20] James A. Van Allen, Carl E. McIlwain, and George H. Ludwig, "Satellite Observations of Electrons Artificially Injected into the Geomagnetic Field," *Journal of Geophysical Research*, 64 (1959): pp. 877-891.

[21] *Scientific Effects of Artificially Introduced Radiations at High Altitudes—A Symposium*, Seven observational and theoretical papers by various authors, in the *Journal of Geophysical Research*, 64 (1959): pp. 865-938.

ing a full geometrical description of the locus of trapping by "labeled" particles, and in establishing an upper limit on the diffusion coefficient of the constituent electrons.[22]

During the remainder of 1958 there were three U.S. Moon shots, *Pioneers I, II,* and *III,* launched on 11 October, 8 November, and 6 December, respectively. The most successful of these was *Pioneer III,* which did not achieve Earth escape velocity but had its apogee at 17 R_E (geocentric). Radiation data were obtained along a full radial traversal of the radiation belt region for the first time—in fact, along two different paths, one outbound away from the Earth and one inbound toward the Earth. It was found that the outer boundary of geomagnetic trapping was at a radial distance of about 10 R_E. In addition, the combination of data from *Pioneer III* with the low altitude data from *Explorer IV* revealed for the first time the full geometrical form of the two distinctly different radiation belts, an *inner* and an *outer* one (Figure 6).[23]

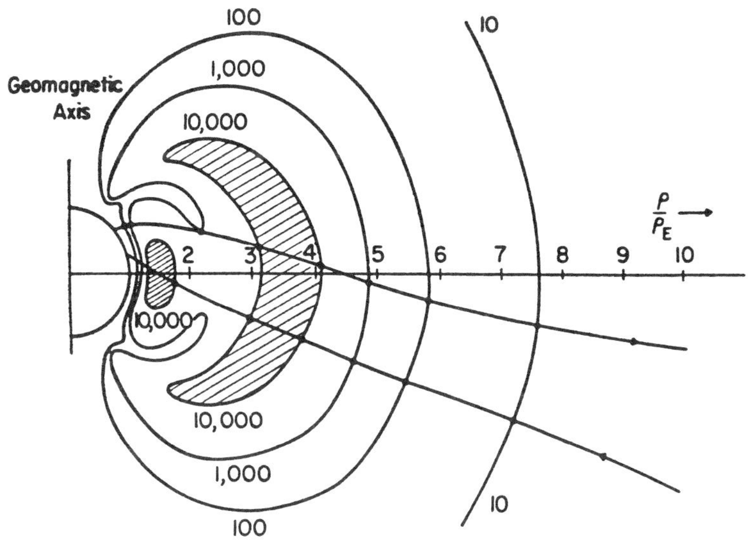

Figure 6 Intensity structure of the trapped radiation around the Earth. The diagram is a geomagnetic meridian section of a three-dimensional figure of revolution around the geomagnetic axis. Contours of constant intensity are labeled with number 10, 100, 1000, and 10,000. These numbers are the true counting rates of an Anton 302 Geiger tube carried by *Explorer IV* and *Pioneer III.* The linear scale of the diagram is relative to the radius of the Earth, 6371 km. The outbound and inbound legs of the trajectory of *Pioneer III* are shown by the slanting, undulating lines. The intensity structure is a function of detector characteristics (Van Allen and Frank, "Radiation Around the Earth to a Radial Distance of 107,400 km," as corrected in James A. Van Allen, "The Geomagnetically-Trapped Corpuscular Radiation," *Journal of geophysical Research,* 64 (1959): pp. 1683-1689.

[22] James A. Van Allen, Carl E. McIlwain, and George H. Ludwig, "Satellite Observations of Electrons Artificially Injected into the Geomagnetic Field."

[23] James A. Van Allen and L. A. Frank, "Radiation Around the Earth to a Radial Distance of 107,400 km," *Nature,* 183 (1959): pp. 430-434.

Pioneer III completed the discovery phase in the study of the Earth's radiation belts, which were subsequently recognized as the distributions of the quite energetic components of the much greater population of (principally) lower energy particles constituting the full plasma physical system, later called the magnetosphere.

Chapter 7

Reaction Motors, Inc.: A Corporate History, 1941-1958[1]

Part I, Institutional Developments

Frederick I. Ordway, III[2] and Frank H. Winter[3]

America's first enterprise devoted to the commercialization of the rocket engine was created 50 years ago in the small town of Pompton Plains, New Jersey. Named appropriately, Reaction Motors, Inc., it came into being less than a fortnight after the United States entered World War II. Its four founders (Figure 1) were rocket enthusiasts who for years had been intimately connected with the American Rocket Society (ARS). All were convinced of the military and business potential of the rocket in the expanding world conflict.

Founding the Enterprise

Enthusiasm was bolstered in 1940, when two of their number, H. Franklin Pierce and John Shesta, received a request from promoters anxious to obtain their help in the development of war rockets. Declining the offer—to Shesta, their actions "did not in-

[1] Presented at the Sixteenth History Symposium of the International Academy of Astronautics, Paris, France, 1982.

[2] The Space and Rocket Center, Huntsville, Alabama, U.S.A.; formerly of Engineering Division, Reaction Motors, Inc.

[3] Department of Space History, National Air and Space Museum, Smithsonian Institution, Washington, D.C., U.S.A.

spire confidence; I was rather luke-warm to the whole idea"—they and their colleagues, Lovell Lawrence, Jr. and James H. Wyld, began to explore ways of attracting the U.S. military to the potential of their own liquid propellant rocket engine.[4]

Figure 1 The four founders of Reaction Motors, Inc. Shown here are: Lovell Lawrence, Jr., first president (top left); Hugh Franklin Pierce (top right); John Shesta (bottom left); and James Hart Wyld (bottom right). Courtesy of Reaction Motors, Inc.

---

[4] Drafts of this paper were reviewed by Laurence P. Heath, Charles W. Newhall, Jr., M. E. "Bud" Parker, John Shesta, and Robertson Youngquist, to whom special acknowledgment is given. Thanks are extended to Thiokol (later, Morton-Thiokol) for allowing access to RMI files located in the Thiokol Headquarters building, Newton, Pennsylvania. (These files were since transferred to Chicago Morton-Thiokol Headquarters and to the RMI archives at The Space and Rocket Center, Huntsville, Alabama.) Illustrations were supplied by Laurence P. Heath, Kew Kyte, the Frederick I. Ordway Collection at The Space and Rocket Center, the National Air and Space Museum of the Smithsonian Institution (Shesta, Lawrence and Wyld collections), John Shesta, and Thiokol.

The four men decided that the best approach would be to demonstrate to the military the viability of the 445 Newton (100 lb) thrust regeneratively-cooled rocket motor, which had been developed and tested by Wyld for several years. The time was appropriate, for ARS testing had accelerated during the summer of 1941 at a site selected by Lawrence in New Jersey. There, Wyld and his colleagues repeatedly fired the 445 N (100 lb) unit for up to 40 seconds duration. Figure 2 shows this engine and Figure 3 illustrates another RMI cylinder with variable thrust from 700 to 2,000 pounds thrust developed later during the 1940s and 1950s.

Figure 2 RMI's 100-lb. thrust rocket cylinder developed under the company's first contract with the U.S. Navy. Courtesy of Reaction Motors, Inc.

Heartened by the success of the motor, Lawrence got in touch with George Lewis, then director of aeronautical research at the National Advisory Committee for Aeronautics (NACA). Lewis, in turn, solicited the Navy's Bureau of Aeronautics (BuAer) in Washington. Interest was aroused as the agency began studying the potential of the Wyld engine.

During this same period, Lawrence, Pierce, Shesta, and Wyld were meeting to consider the idea of forming a company. The name Reaction Motors was put forward by Wyld, who saw it as a kind of General Motors of the rocket world. "The analogy was not very close," he later admitted, "as we had scarcely two nickels to rub together, and our plant consisted mostly of half of the upper floor of John Shesta's brother-in-law's garage in North Arlington, New Jersey, which was about as large as a rather spacious outhouse."[5]

---

[5] From "A Biographical Sketch of the Rocket Career of James Wyld," prepared by Wyld and residing in the RMI files at the Thiokol Corporation, Newton, Pennsylvania. The five-page document was prepared ca. 1951. The contract referred to was for $39,000 and was awarded by BuAer.

The men persisted in their aim of obtaining support from BuAer and, according to Wyld, "Within a few weeks [of earlier contact] Lawrence had succeeded in wrangling a visit [in November 1941] from a U.S. Navy Representative to witness a test run of the regenerative motor—again in a hidden spot in the woods near Wanaque. It was a good run, and he departed quite enthusiastic." The Japanese attack on Pearl Harbor occurred shortly afterward, followed almost immediately by RMI's first Navy contract. "Very soon," continued Wyld, "we were installed in our own shop and working like beavers on our first experimental motors for the Navy."[6]

Figure 3 Early 700-2,000 lb variable thrust engine. Courtesy of Reaction Motors, Inc.

[6] *Ibid.*

The initial meeting of the new company's incorporators was held on 16 December 1941 in New York City. The organizers of record were Harry B. Davis, Charles N. Caldwell, Jr., and Wilson D. Smith, whom Lawrence's lawyer "dug up," in Shesta's words, to serve as "first stage dummies." On the same day, the certificate of incorporation of the company was filed with the secretary of state of New Jersey; and two days later a certified copy was recorded in the office of the county clerk, Hudson City, New Jersey. At the 16 December meeting, transfers of stock subscriptions took place from Davis, Caldwell, and Smith to "second stage" dummy incorporators H. W. Reese (Shesta's brother-in-law), B. D. Lawrence (one of Lovell's brothers), and Beverly Blizzard (Wyld's brother-in-law). As Shesta explains, "our lawyer felt that if there were ever an investigation of RMI, the opposing lawyer would first check out the first stage incorporators, and then the second, and would finally give up!" He would later remark that "I could never understand the reason for this silly charade . . . concocted by the lawyer . . . but that is the way it was."

At 1 p.m. on 16 December the just constituted Board of Directors met in New York. The first order of business was for the corporation to receive one liquid-propellant rocket engine valued at $5,000 from Lovell Lawrence, Jr., John Shesta, H. Franklin Pierce and James H. Wyld. Stock transfers were then effected, followed by the resignations of Blizzard, Reese, and B. D. Lawrence and their replacement by Lovell, Lawrence, Shesta, Pierce, and Wyld. This done, a 15 minute recess was taken. Upon reconvening, officers were elected and treasurer Shesta was authorized to open a bank account in RMI's name. Finally, the board authorized RMI's president Lawrence to enter into negotiations with the Navy "to sell to said department the Liquid Fuel Rocket Motor for $5,000.00 and to offer the services (without charge) of the company's engineer (Wyld) to install and instruct in the use of same; conditioned, however, that said New Department commission this Corporation to build a similar motor capable of a thrust reaction of 1,000 lbs., for the sum of $20,000.00, delivery to be made within five (5) months of the signing of the contract."[7]

According to John Shesta, the Navy set down the following obligations to be fulfilled by RMI within 180 days:

1. Deliver the existing Wyld motor to the Navy.
2. Develop and demonstrate a similar 445 N (100 lb) thrust motor operating on aviation gasoline and liquid oxygen.
3. Develop a 4450 N (1,000 lb) thrust motor using the same propellants.
4. Demonstrate repeated starts and throttling of the engine down to half thrust.[8]

They got the Wyld motor off to the Navy on schedule and received a welcome $5,000. "With this money," recalled Shesta, "we purchased a one ton truck, rented an

[7] From the Minutes of the First Meeting of Incorporators and the Minutes of the First Meeting of the Board of Directors, RMI files, Thiokol Corporation, Newton, Pennsylvania.

[8] John Shesta, "Reaction Motors, Incorporated—First Large Scale American Rocket Company: A Memoir," paper presented at the 29th Congress of the International Astronautical Congress, Dubrovnik, Yugoslavia, October 1978, copy in possession of authors.

old building and set up shop in Pompton Lakes, New Jersey. We assigned ourselves nominal salaries and went to work."[9] Among the first to join the four founders were Louis Arata, Leslie Collins, Kirk Fischer, and Charles H. Dimmick.

The Four Founders

Who were the founders? What did they have in common? And what brought them together to work "like beavers" under marginal conditions on an only slightly understood technology? First and foremost, all were amateur rocket experimenters who had been working together since the mid-1930s under the auspices of the ARS. Each was technically trained and had worked on engineering and related projects. And the four men shared an entrepreneurial spirit.[10]

Lovell Lawrence, Jr., president, was a graduate of Montclair State College in New Jersey. Prior to RMI, he had worked for seven years at the International Business Machines Corporation as assistant to the chief engineer, Radiotype Division. There, he designed and supervised the installation of automatic radio-writing equipment for the U.S. government. Lovell would remain with RMI until 1951 when, because of gradual loss of administrative and technical authority, he left the company. He died in 1971.

H. Franklin Pierce served as vice president and head of test engineering. As early as 1929, he had become interested in rockets and had constructed an engine propelled by alcohol and liquid oxygen. With no built-in igniter, it had to be fired up by a flaming torch at the end of a long stick! During the same period in Germany, of course, Willy Ley, Wernher von Braun, and other German Rocket Society (VfR) experimenters were employing equally primitive techniques. Prior to joining RMI, Pierce served four years in the U.S. Navy as a machinist and later worked in a similar capacity with New York City's Interborough Rapid Transit subway system. He stayed with RMI until 1947, when, in Shesta's words, he "more or less got bored with rockets," sold his stock, and departed for Orosi, California, to run a citrus farm.[11] Dimmick learned that Pierce had failed in the citrus business by 1960 and had taken a position at the Douglas Aircraft Company.[12] No later information is available on Pierce.

The positions of treasurer and director of research and engineering were occupied by John Shesta. A graduate and post-graduate of Columbia University and later a civil engineering instructor there, he resigned as chief engineer of the Z-Flex Piston Ring Corporation to join RMI. Shesta was a well-known designer and experimenter at the ARS. When the great economic depression threw him out of work, he said he was

[9] *Ibid.* Shesta reported the four men had but two small lathes, a drill press, an assortment of small tools and an oxy-actelyne welding outfit as their company's equipment at the time.

[10] At one time or another, all had been or would be ARS presidents: Shesta, 1936; Pierce, 1940-1941; Wyld, 1944-1945, and Lawrence, 1946.

[11] John Shesta to Frederick I. Ordway III, 2 August 1982.

[12] Interview with Dimmick by Ordway, RMI reunion, Lake Mohawk, New Jersey, 10 October 1982.

thankful: "my job was beginning to interfere with my real work," e.g., testing rockets with his ARS colleagues.[13] He later became vice president, a position he held until 1952, when he resigned. Early the following year, he also left the company's board of directors. He then retired from technical work and moved to Morrisville, Vermont, where he died in 1987.

James H. Wyld accomplished his undergraduate and post-graduate work at Princeton University. Before joining RMI as secretary and chief research engineer, he was a design engineer at the Rahm Instrument Corporation. His 445 N (100 lb) thrust, liquid oxygen-alcohol rocket motor was, in the inventor's words, "the first American design to apply generative cooling to all parts of the motor." Wyld died on 3 December 1953 while on sick leave from the company when he was only 41 years of age.[14]

During the ensuing years of RMI's growth, many other professionals joined the company's ranks. On the engineering and scientific sides, among the better known were William P. Munger, top-flight turbopump designer and long-time chief engineer; project engineers M. E. "Bud" Parker, Edward A. Neu, Jr., Bernard Pearlman, Arthur Sherman, Robertson Youngquist, Whitney Nesbitt, Haakon Pederson, Robert Seaman, Walter Borcherdt, Delacy F. Ferris, Harold Davies, Neil O'Rourke and Harry W. Burdett, Jr. Other key people included William R. Wright, Jr., patents engineer; Paul F. Winternitz, expert in physical chemistry and at one time or another director of research and of laboratories; Robert L. Wehrli, research physicist; and Harry B. Horne, chemical engineer and engineering manager. Joseph W. Mollek was RMI's top manufacturing expert; Warren P. Turner was its applications engineering and contracts head; and Lawrence P. Heath was its government relations coordinator. On the financial side Henry H. "Mike" Michaels played a key role.

The Jato Endeavor

To get their program under way, RMI's four founders quickly focused attention on meeting the requirements for a 4450 N (1,000 lb) thrust engine, which was to be employed by the Navy to assist large, heavily-laden flying boats during takeoff. According to Shesta, by the end of November 1943, he and his associates were ready to start work with the Navy in Annapolis. There, a nitric acid based rocket program was underway at the Naval Engineering Experiment Station (NEES). Dr. Robert H. Goddard was working at NEES on pumps and turbines in a small building assigned to him. Not very exciting, perhaps, but, as Lawrence later pointed out, "we had to hang our hat on something that would sound practical to the greatest doubting Thomas. The words rocket and missile had to be handled with great tenderness." Hang their hat, or rather hats, they did.

[13] Shesta quote, *Boston Herndon*, "Rocket Genius with Big Ideas," *Coronet*, April 1953, p. 26.

[14] James H. Wyld, "The Liquid Propellant Rocket Engine," *Mechanical Engineering*, June 1947, p. 5. Just before his death Wyld had been on a year-long loan to Oak Ridge, Tennessee, where he investigated the potential of nuclear fission for rocket propulsion.

The early RMI rocket motors were much like the regeneratively cooled unit Wyld had developed in 1938. Tests and improvements registered during the first half of 1942 helped bring thrust ratings up, with the Navy-contracted 4450 N (1,000 lb) level being reached by November. A full 15,130 N (3,400 lb) of thrust was attained by May 1943, in response to a second Navy contract that called for at least 13,350 N (3,000 lb). All the while, company personnel strength was increasing from the original four in 1941 to nearly 20 as 1942 ended. Figure 4 shows the engine mounted in a Navy PBM flying boat in early 1944 at Annapolis, Maryland, with key personnel standing in front.

Early Expansion

Expanded responsibilities and increased personnel put pressure on office, shop and test facilities. Things were primitive indeed at first. The Pompton Plains garage area first occupied by the company was 22.6 m^2 (800 ft^2) and was partitioned. One section served for administration and drafting, the other as machine shop and laboratory. Looking back from the vantage point of ten years, the founders felt that:

> In view of the company's limited requirements, the plant was efficient. When one of the four officer-employees received the nucleus of an idea, he reduced it to a design in the office drafting room, then donned a set of overalls and machined and assembled the unit. When it was completed, he tested it by pointing the proper end out of the back door and firing it into the New Jersey countryside.[15]

That kind of operation sufficed during 1942, but by year's end it had become clear that expansion was in order. Consequently, the following January, the new firm moved into a three-story frame building in Pompton Plains (Figures 5 and 6). The 1077.6 m^2 (11,600 ft^2) structure had previously been a silver factory, dating from the late nineteenth century and then "The Silver Circle" night club. Lawrence, Shesta, Pierce and other early employees of RMI stood in front of the Pompton Plains building in Figure 7 and in the parking lot is the old RMI Ford truck (Figure 8).

Leslie Collins, an early RMI employee, recalls that the move into the stucco exterior structure "went off very smoothly and all hands were in there pitching. There were the President, Vice President, Treasurer, Secretary and Foreman (Collins himself), and all the other employees (three), wiring up machines, drilling holes in concrete, setting up the engineering department and stock room. By Monday, 17 January 1943, the new plant was all ready to go."[16]

In addition to the old silver factory, RMI took advantage of the 1.4 hectares (3.5 acres) around it to erect three stands. One was used for testing liquid propellant rocket motors with thrust ratings as high as 6670 N (1,500 lb) operating on liquid oxygen oxidizer and both gasoline and alcohol fuels. The second tested multi-motor units typified by RMI's 4-chamber, 26,700 N (6,000 lb) thrust aircraft engines. And the third was

---

[15] "Hobby Discussion Led to RMI Birth," *The RMI Rocket*, December 1951, p. 2.

[16] *Ibid.*, p. 6.

employed in resonating-type, duct-engine testing. The company also maintained their original stand at nearby Franklin Lakes (Figure 9) to test liquid oxygen and alcohol and gasoline motors at thrust ratings up to 13,350 N (3,000 lb). Figure 10 is a photograph of an early test stand there, and Figure 11 shows the installations with John Shesta kneeling by the 350-pound thrust cylinder.

Figure 4 Reaction Motors, Inc. and U.S. Navy crew who worked on Reaction Motors first JATOs, 1944 by a PBM seaplane at Annapolis, Maryland. Left to right: Lovell Lawrence, Jr.; Frank Muth, head draftsman, Reaction Motors; C. E. Bellais, CWO; Lieut. William Gore; H. F. Pierce in back row, behind Gore; Les Collins of Reaction Motors; and John Shesta. The engine is a 3,000 lb thrust single cylinder model. This was Reaction Motors first 3,000 lb motor on a plane, PBM-3 No. 6586. Courtesy of Reaction Motors, Inc.

Figure 5 The second home of Reaction Motors, Inc. (1943-1946), at Pompton Plains, New Jersey, which opened January 17, 1943. Rocket static stands were in adjacent grounds. Courtesy of Reaction Motors, Inc.

Figure 6 A closeup of the old silver factory, second home of Reaction Motors, Inc., Pompton Plains, New Jersey. Courtesy of Reaction Motors, Inc.

Figure 7 Lovell Lawrence, Jr., John Shesta, H. F. Pierce, and other early employees of Reaction Motors, Inc., standing in front of the Pomton Plains building. Courtesy of Reaction Motors, Inc.

Figure 8 Reaction Motors, Inc., Ford truck purchased with first earnings of the company in front of the old silver factory, Pompton Plains, New Jersey. It was used during the World War II years to transport rocket supplies and equipment. Courtesy of Reaction Motors, Inc.

Figure 9 Reaction Motors, Inc. test area, Franklin Lakes, New Jersey, in September 1942, before the company built its first large stand. Note the viewing ports in the primitive blockhouse. Courtesy of Reaction Motors, Inc.

Figure 10 An early test stand with a small rocket on it, Franklin Lakes, New Jersey. Courtesy of Reaction Motors, Inc.

Figure 11 John Shesta kneeling in front of a 350-pound thrust rocket cylinder in July 1944, Franklin Lakes, New Jersey. Courtesy of Reaction Motors, Inc.

Rocket Companies Make Poor Neighbors

Reaction Motors' location inevitably led to complaints; after all, rocket motors are not test-fired in silence. As an engineering reporter observed, "From a window of its single building one could toss a rock to the limits of its crowded testing area."[17] A petition of irate citizens was soon drawn up alleging that vibrating noises and explosive detonations were "damaging the properties and health of the signers and their families." Local committeeman Emmet Petrin is reported to have said, "It's pretty evident that Reaction is hurting our town; and the other members of the committee, speaking unofficially, agree they ought not to be in a residential area." A young Riverdale housewife lamented, "When those tests go off, I'm scared out of my wits. The baby cries and the house shivers. Reaction ought to go some place, far, far away.[18]

A week later the *Newark Evening News* reported on a protest by the Pequannock Township Committee in which it was stated that RMI may go "out of town, out of business or out to the Pacific." It went on to add that Pequannock residents had "prayed for the necessary relief" from RMI's experiments, which "are damaging the properties and health of the signers [of a written protest to the township] and their families." It was further reported that Police Chief Walter Sweetman had served notice on the company that its application of 23 August to store explosives had been denied.[19] The license was officially revoked by him on 25 October.

Move to Lake Denmark

The result of all this was a mid-1946 move to a location at the U.S. Naval Ammunition Supply Depot at Lake Denmark, New Jersey, close to Dover. The change was particularly advantageous for the company's test operations, which could now take place at the 10 hectare (25 acre) site with acoustic shielding afforded by wooden hills.[20] Larry Heath recalled how the site was selected. One day, Lovell Lawrence said, "Heath, I want you to find us another plant location. I've got to have a remote location. I don't care where, in the mountains, plains or woods, but its got to be remote. You can locate it anywhere in the Untied States as long as it's within 15 minutes of my new house [in

[17] John O. Davies, Jr., "Rocket Business Soars," *Newark Evening News*, 18 September 1946.

[18] "Residents Cite Damage Done by Tests of Jet Propulsion," *Newark Evening News*, 2 October 1946.

[19] "Pequannock Bans Jet Motors Testing," *Newark Evening News*, 10 October 1945.

[20] Established in 1891 by Act of Congress, the depot occupied a tract of land comprising 315 acres that had formerly been part of the United States Powder Depot (later to be known as Picatinny Arsenal). An additional 78 acres were purchased in 1902; and, in 1917 by Presidential Proclamation, 67 more were acquired. In 1948, the Naval Aeronautical Rocket Laboratory was established on the site with the mission to develop and evaluate liquid-propellant rocket engines and their components. On 1 April 1950, the Naval Ammunition Supply Depot's name was officially changed to the Naval Air Rocket Test Station. Simultaneously, cognizance would be transferred from the Bureau of Ordnance to the Bureau of Aeronautics.

Denville, New Jersey]!" With the help of the Navy, "the Depot popped up, right next door . . . with its long, empty buildings which to us looked like a palace."[21]

Two large buildings in the old ordnance battalion area at Lake Denmark housed RMI's engine design and model construction and production facilities. Smaller structures served for administration, security, and other supporting services. Testing took place in a remote area on a narrow dirt road between the depot and Hibernia. That the company was pleased with its new location is attested to in its first annual report:

> The location in a natural "bowl" of surrounding hills insulated by thousands of acres of government property acquired specifically for ammunition storage and testing provides one of the best areas on the Eastern seaboard for rocket engine testing, yet within easy distance of an ample supply of skilled labor and test fuels. Following the move to Lake Denmark the construction of substantial testing facilities was started on Navy property and today the company operates one of the largest and best equipped liquid rocket engine testing facilities in the country. As a safeguard for future testing requirements, the Company has acquired a parcel of 237 acres of land just north of the Navy sponsored test area.[22]

Figure 12 shows two rocket engine tests underway.

Figure 12 Early Reaction Motors, Inc. small rocket engine being tested in new test stand, Lake Denmark, New Jersey. Courtesy of Reaction Motors, Inc.

---

[21] Laurence P. Heath interview by Frederick I. Ordway III, Morrisville, Vermont, 5 July 1982.

[22] From the *First Annual Report of Reaction Motors, Inc.*, for the period ending 31 December 1948 and covering the years 1947 and 1948.

Three years after moving from Pompton Plains to Lake Denmark, RMI shifted its administrative, personnel, finance, and manufacturing functions five miles away to Rockaway. The move was celebrated on 6 December 1949, when the new facilities were inspected by a distinguished group that included Under-Secretary of the Navy, Daniel A. Kimball; Rear Admiral Lloyd Harrison, BuAer assistant chief of design and engineering; Rear Admiral Calvin M. Bolster, BuAer's assistant chief for Research and Development (R&D); Commander Robert C. Truax, Naval Aeronautic Rocket Laboratory; Air Force Procurement head for the New York region, Brigadier General Arthur Thomas; Colonel J. W. Sessums, Jr., deputy chief of R&D at the Air Force Material Command at Wright Field; and Laurence S. Rockefeller, among others. Figures 13, 14, 15, 16, and 17 are Rockaway buildings.

Figure 13 Interior view of Reaction Motors, Inc. workshop, Rockaway, New Jersey. Courtesy of Reaction Motors, Inc.

Figure 14 Recent view of Reaction Motors, Inc. old engineering building, Rockaway, New Jersey. Engineering operations were conducted on the top floor, manufacturing on the lower floor. Courtesy of Frederick I. Ordway III.

Figure 15 Reaction Motors, Inc., pump and hydraulics laboratory on Stickle Avenue, Rockaway, New Jersey, looking west (recent photograph). Courtesy of Frederick I. Ordway III.

Figure 16 Recent overall view of Reaction Motors, Inc., facilities, Rockaway, New Jersey (looking east). Foreground: first floor, manufacturing, second floor, design offices. In the background: first floor, engineering records, second floor, project engineering. Courtesy of Frederick I. Ordway III.

Figure 17 Closeup view of the building in Rockaway, New Jersey originally used by Reaction Motors, Inc., on Oak and Stickle, looking north. Records were kept on the first floor and design and engineering offices on the second floor. Courtesy of Frederick I. Ordway III.

Six years later, on 7 October 1955, the company formally dedicated a new and expanded plant at Denville consisting of three main structures; one for engineering and research, one for administration, and one for manufacturing. Utility and plant services were housed in separate buildings. At the same time, RMI's test facilities at Lake Denmark were modernized and expanded (Figure 18).

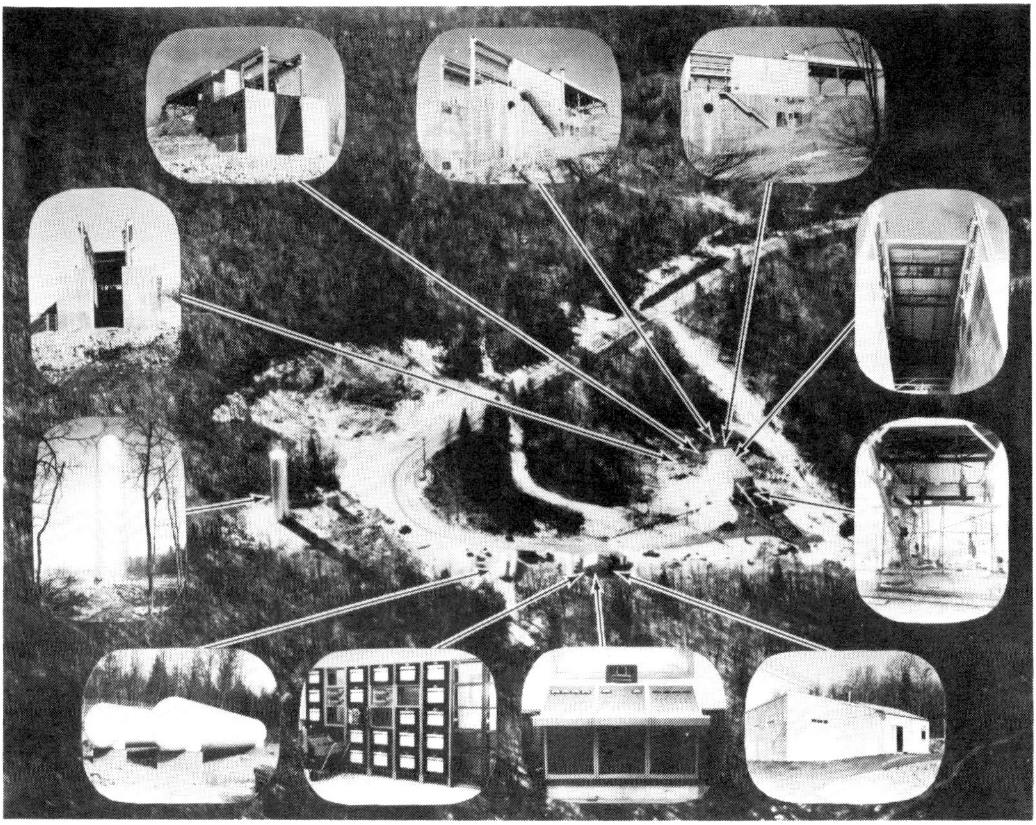

Figure 18 Composite view of the Reaction Motors, Inc. R-2 test area, Lake Denmark, New Jersey, November 1956. This was the main test complex for the XLR-99. Courtesy of Reaction Motors, Inc.

Each of these moves, of course, reflected other company growth. By 1943, the number of employees exceeded 20, reached 55 in 1945, 120 in mid-1946, and 473 in 1947. This growth put an acute strain on RMI finances; while the company was technically prospering, financially it was going down hill. After posting a profit of $643 during 1942 (its initial full year of operation), it earned nothing the next four.[23] At the end of 1946, the firm's net worth was minus $100,000, but by mid-1947, it owed creditors some $600,000. RMI's problem was simple: too many fixed price contracts had become unprofitable when development costs exceeded estimates. Unfortunately, extended production orders were few (Figure 19).

Figure 19 A production line view of the CML2N Gorgon 350 pound thrust cylinder assemblies, left, and Gorgon tanks, right. Courtesy of Reaction Motors, Inc.

[23] This and related information was obtained by Ordway working with the RMI files at Thiokol Corporation Headquarters in Newtown, Pennsylvania, as well as during discussions there with T. W. Brennan; William R. Wright, Jr.; J. T. "Gus" Caton; and T. G. Tumelty on 15 February 1982; during interview with M. E. "Bud" Parker at Thiokol's Huntsville, Alabama office on 26 March 1982; during interviews with John Shesta and Laurence P. Heath in Morrisville, Vermont, 3-6 July 1982; during interviews with Robertson Youngquist in Washington D.C., 27 August and 29 December 1982; and with Charles H. Dimmick and other former RMI employees, Lake Mohawk, New Jersey, 10 October 1982. Statistics were taken from RMI annual reports, beginning with the "First Annual Report of Reaction Motors, Inc., Dover, New Jersey, for the Period Ended December 31, 1948:" (Lake Denmark was officially in Dover) and extending to its last corporate annual report for the year 1957. The first report provided financial and other data for the years 1947 and 1948. In 1959, the company was absorbed by the Thiokol Chemical Corporation and hence issued no separate annual report after that event.

The Rockefeller Connection

Fortunately, at this critical time, Laurence S. Rockefeller came to the rescue. After many months of study and negotiation with RMI officials, Navy officers, and banking executives, he agreed to invest in the company. Charles Watson Newhall, Jr., then of the Rockefeller staff, recalled how the association worked out.[24]

It was during the summer of 1947 that Newhall first became involved with RMI. Later, as RMI's plight worsened, the Navy contacted Rockefeller saying, in effect, that it had a little company out in New Jersey that is making important contributions to Navy and Air Force rocket programs and it is most anxious to keep it alive. How about taking another look?[25]

"This," according to Newhall, "stimulated Laurence's interest. Likewise, it was a 'far out' new field of technical endeavor which was of interest to his new venture capital activities. In this instance, immediate financial success was of secondary consideration—helping the defense effort and pursuing a new technical field were primary." Newhall says he never learned who made the call, but suspects it was someone in Admiral Lloyd Harrison's contract group.[26]

Rockefeller put Newhall in charge of investigating RMI to see if some sort of agreement could be made. Newhall worked closely with Lovell Lawrence, Jr., president; Robert Lawrence, accounting; and Jack Pethick, treasurer. Newhall discovered that as of 31 August 1947 there was a working capital deficit in excess of $250,000!

The agreement between Rockefeller and RMI was signed 10 days later; and, according to Newhall, involved the purchase $2000,000 principal amount of 3-year, 4 percent notes with detachable warrants for the purchase of common stock. In March 1948, this was increased by $100,000. By December of that same year, RMI's financial condition had improved to the point that it could increase its line of revolving credit with Chase and the Paterson banks from $315,000 to $600,000.

"As soon as the agreement was signed," recalled Newhall, "Laurence asked me to move out to RMI and 'hold the fort' until they selected someone as general manager. For the next six weeks, I interviewed numerous candidates sent out by the 'New York office'—also showed them the temporary facilities at Lake Denmark. One afternoon,

---

[24] Charles Watson Newhall to Frederick I. Ordway III, 3 April, 23 May, and 5 November 1982.

[25] Laurence was the "gadgeteer" of the Rockefeller family and for years had been investing in technology-oriented ventures. In the middle 1930s, he had backed World War I ace Eddie Rickenbacker's fledgling Eastern Air Lines; and, in later years, moved into airplane design, electronics and, with RMI, rocket engines. J. S. McDonnell, Jr., of St. Louis received $475,000 of Rockefeller money, most of it Laurence's, to finance his McDonnell Aircraft Corporation. Among the companies receiving Rockefeller venture capital during the post World Was II period were Piasecki Helicopter Company; the Laboratory for Electronics, Inc.; and Universal Business Machines, Inc. "What we are trying to do," explained Laurence at the time, "is to clothe engineering and creative brains with financial and managerial strength." He added, "Instead of holding capital back until a field is proved safe, we are undertaking pioneering projects that, with proper backing, will encourage sound scientific and economic progress in new fields where there will be tremendous future development." Cited by Joe Alex Morris in "The Rockefellers," *The Saturday Evening Post*, 13 January 1951, p. 116.

[26] Newhall to Ordway.

Laurence called and said that he and 'Mac' [J. S. McDonnell of McDonnell Aircraft Co.] would like to come out and see me, 'J.S.' had been working closely with Laurence to help him with RMI."

Newhall remembered their settling into his small office and announcing that they had made their selection of general manager, "and it was me. You could have knocked me over with a feather," he added, "since I didn't know I was under consideration. I was pleased, because I had become very interested in the company and its people."[27]

With the fresh injection of capital, the financial health of RMI improved. Employment increased, new equipment was installed, and business forecasts became optimistic. "In its relatively short life of seven years," a RMI report boasted, "the company has demonstrated its capabilities as one of the prime sources of liquid rocket power plants and accessories for the Services."[28] Examples of the company's achievements at the time are summarized in Table 1.

Table 1

ACHIEVEMENTS OF REACTION MOTORS DURING FIRST SEVEN YEARS

1. First American Corporation dedicated solely to the development of liquid rocket engines and accessory equipment.
2. First development of regeneratively cooled liquid rocket engine generating 1,000 pounds of thrust using liquid oxygen and alcohol propellants.
3. First regeneratively cooled liquid rocket engine generating 3,000 pounds of thrust using liquid oxygen and gasoline propellants; powered first successful liquid rocket JATO unit in PBM3C flying boat.
4. First rocket engine to propel a piloted aircraft faster than the speed of sound; Air Force Xs-1. Many supersonic flights accomplished; impressive field record established by the four-chamber 6000-pound thrust engine.
5. First practical U.S.-developed turbine pump system for use with rocket engines. Basic design used with four engine models, ranging from 1,500 pounds of thrust; driven by decomposing hydrogen peroxide in presence of catalyst.
6. First successful U.S.-developed turbine pump-driven power-plant for guided missile application; Air Force MC-774. Unique thrust chamber swiveling feature to steer missile on its course successfully demonstrated.
7. First successful U.S.-developed turbine pump-driven rocket engine for piloted aircraft; Navy D558-2 then undergoing initial flight tests with 6,000-pound thrust engine.

[27] Newhall would remain as executive vice president and (to 1951 general manager) until 1953, when Rockefeller sent him to Flight Refueling, Inc. of Danbury, Connecticut, to study it as a possible RMI-Rockefeller joint investment. In January, a controlling interest in the firm was acquired from its British owner, Flight Refueling, Ltd. of Dorset. And in April, Newhall was made president and treasurer of Flight Refueling, Inc., which was relocating to new facilities in Baltimore, Maryland.

[28] Section C-2(e), "History and Accomplishments," Report ANC-1062, 1948, RMI archives, Thiokol Corporation, Newtown, Pennsylvania (unpaginated).

8. First development of 20,000-pound thrust, turbine-driven liquid oxygen-alcohol rocket engine for Navy's Viking high-altitude sounding rocket.
9. The development of 350-pound thrust, two-cylinder, acid-aniline engines for Navy Lark; several hundred delivered and flight tested.
10. Construction of fuel, oxidizer, and pressurizing service carts for Lark engines.
11. First variable thrust, constant chamber pressure rocket engine; minimum thrust limit of 20 percent of rated 2,000 pounds.
12. Development of automatic and manual systems for varying thrust by throttling of propellants.
13. First liquid rocket engine application to rotary wing craft; 50-pound thrust, liquid oxygen-alcohol thrust chamber used.
14. Investigations and tests on resojet-type engines.
15. Application of rocket power to boats.
16. Research, development and testing of high-energy propellants, e.g., diborane and hydrazine as well as ammonia, turpentine, and furfuryl alcohol.
17. Development of means for determining temperature of rocket jets by calculation and observation.
18. Investigations on use of liquid oxygen, white fuming nitric acid, JP3, and gasoline as propellants in systems of 8,000 and 16,000 pounds of thrust, including thrust cylinders, propellant pumps, gas generators and controls.
19. Development of a complete line of accessories for rocket engines.
20. Development of special tools and equipment for constructing and checking rocket engine components and accessories.
21. Design construction and operation of liquid rocket test areas, including test equipment and specialized instruments. Thrust capacity to 25,000 pounds (site at time contained nine operating test stands with another four under construction).

Company growth continued; by 1951, 642 employees were on the payroll. Meanwhile, in early 1950, Raymond W. Young, Jr. arrived as vice president of engineering, following a career of 24 years at Wright Aeronautical Corporation. In October 1951, he replaced Lovell Laurence, Jr., as president, who subsequently resigned and took a position with Chrysler's missile operations. And three months later, in January 1952, Newhall reported to the New York Society of Security Analysts that RMI had earned about $1.25 a share on 109,000 common shares in 1951, and that sales had reached some $4.5 million (up a million dollars from the previous year). In December 1952, he announced a $4 million expansion program for the 11-year old firm.

Olin-Mathieson Arrives on the Scene

In late November 1953, slightly more than six years after the RMI-Rockefeller agreement was negotiated, the Mathieson Chemical Corporation bought 50 percent of RMI's outstanding stock and hence a controlling interest in the company. The following

May, Olin Industries, Inc. and Mathieson merged into the Olin Mathieson Chemical Corporation. The 1955 slate of officers reflected the changed control of the 14-year old RMI (Table 2).

Table 2

DIRECTORS AND OFFICERS AT RMI 1949, 1952, 1958.

Directors

1949

Lovell Lawrence, Jr., president RMI

Randolph B. Marston, associate of Laurence S. Rockefeller, New York City

Charles W. Newhall, Jr., Executive vice president and general manager, RMI

Stuart N. Scott, Root, Ballantine, Harlan, Bushby & Palmer, New York City counsel

John Shesta, Director of research and engineering, RMI

Harper Woodward, associate of Laurence S. Rockefeller, New York City

James H. Wyld, chief research engineer, RMI

1955

Myron B. Gordon, aviation consultant, New York City

Russell Hopkinson, vice president for development, Olin Mathieson Chemical Corporation

Hans A. Klagsbrunn, Kingsbrunn, Hanes and Irwin, Washington, D.C.

Randolph B. Marston (same as 1949)

Henry H. Michaels, J., executive vice president, RMI

Walter F. O'Connell, vice president for finance, Olin Mathieson Chemical Corporation

Dr. Fred Olsen, vice president for research, Olin Mathieson Chemical Corporation

Harry A. Sosnoski, director of operations, Aviation Division, Olin Mathieson Chemical Corporation

Harper Woodward (same as 1949)

Raymond W. Young, president, RMI

1958

William C. Foster, chairman; senior vice president and director, Olin Mathieson Chemical Corporation

Myron B. Gordon (same as 1955)

L. Kermit Herndon, divisional vice president for research and production of the High Energy Fuels Division, Olin Mathieson Chemical Corporation

Hans A. Klagsbrunn (same as 1955)

Randolph B. Marston (same as 1949 and 1955).

Henry H. Michaels, Jr., senior vice president and treasurer, RMI

Harry A. Sosnoski, divisional vice president for customer relations of the High Energy Fuels Division, Olin Mathieson Chemical Corporation

Harper Woodward (same as 1949 and 1955)

Raymond W. Young (same as 1955)

Lovell Lawrence, Jr., president

Charles W. Newhall, Jr., executive vice president and general manager

Henry H. Michaels, Jr., Vice president of finance and administration

John Shesta, director of research and engineering

Harry B. Horne, Jr., assistant to executive vice president and general manager and assistant secretary

Alexander L. Keyes, secretary and general counsel

John A. Pethick, treasurer

Joseph W. Mollek, Jr., manager of manufacturing

Laurence P. Heath, manager of contract administration and service

James H. Wyld, chief research engineer

Howard Moles, chief engineer

Paul Winternitz, director of laboratories

Officers

1948

Raymond W. Young, president

Henry H. Michaels, Jr., executive vice president

Robert S. Kinsey, vice president of operations

Alexander L. Keyes, secretary and counsel

James W. Fay, Jr., treasurer

William P. Munger, manager of engineering and research

David C. Eaton, works manager

Warren P. Turner, manager of customer relations and contracts

David G. Keller, manager of industrial and public relations

Robert M. Lawrence, assistant treasurer

1958

Raymond W. Young, president

Henry H. Michaels, Jr., senior vice president and treasurer

Joseph W. Antonides, vice president and executive assistant to the president

David C. Eaton, vice president for marketing

Fred D. Pfotenhauer, vice president for production

Alexander L. Keyes, secretary and counsel

Robert M. Laurence, assistant treasurer and director of financial division

Delacy F. Ferris, chief engineer

Edmund P. Parke, director of production division

Edward H. Seymour, director of research division

A. Robert Taylor, director of administration division

All the time, RMI was growing. At the end of 1954, the work backlog reached more than $7.5 million and sales climbed to $4.7 million. The next year, sales exceeded 1954's backlog, which in turn had soared to over $9 million. Sales exceeded $16 million in 1956 and almost reached $24.5 million in 1957. Between 1955 and early 1958, the number of employees rose from 669 to 1,639.

The Thiokol Merger

RMI's activities were being closely watched by another expanding company, the Thiokol Chemical Corporation. Founded in 1929 as the first commercial producer of synthetic rubber in the United States, the firm had moved into the solid-propellant rocket field following World War II. Its 1957 sales approximated $30 million. A merger between Thiokol and RMI seemed appropriate to the directors of both companies, as well as to Olin Mathieson and Rockefeller, who between them controlled more than two-thirds of the outstanding RMI stock.[29]

A merger with RMI was proposed to Thiokol stockholders at a meeting in Trenton, New Jersey, on 17 April 1958. They were advised that if the merger were to become effective, "approximately 200,000 [actually, 198,000] additional shares of capital stock of Thiokol Chemical Corporation will be issued to the holders of outstanding stock of RMI." Moreover, the approval by holders of 66-2/3 percent of Thiokol's outstanding capital stock would be required to authorize the proposed merger. Thiokol stockholders were informed by President J. W. Crosby that "All of your directors join me in recommending a vote FOR this proposal." Thiokol stockholders were enthusiastic, voting 777,191 shares for the merger and a mere 3,884 against.[30]

The merger plan dominated the regular meeting of RMI's board of directors held in New York City on 24 January 1958. Two documents were distributed at the meeting dated 24 January; a form of agreement between Thiokol and RMI, and a form of letter agreement between Thiokol and RMI's two principal stockholders, Olin Mathieson and Laurence S. Rockefeller. It was unanimously resolved:

[29] The RMI-Olin Mathieson arrangement was somewhat less than satisfactory, with the latter seeking more management control than RMI president Young was prepared to yield. Consequently, there was little Olin Mathieson resistance to the Thiokol proposal.

[30] Message to stockholders, 17 April 1958, Thiokol Chemical Corporation.

that the proposed Agreement, dated 24 January, 1958, between Thiokol Chemical Corporation and this Company, in the form submitted to this meeting, be and the same hereby is approved, and that the proper officers of the Company be and they hereby are authorized and directed to execute and deliver such Agreement to Thiokol Chemical Corporation, in the name and on behalf of the Company.

William C. Foster, chairman of RMI's board of directors and executive vice president of Olin Mathieson, announced the overwhelming Reaction Motors' stockholder response: 203,802 shares were for the merger with Thiokol and 1,425 shares were opposed. The two companies were officially merged at midnight on 30 April 1958, with RMI becoming the Reaction Motors Division of the Thiokol Chemical Corporation. The final phase of RMI's history was about to begin.[31]

[31] The history of Reaction Motors Division of the Thiokol Chemical Corporation appears in Chapters 11 and 12 of Volume 12, *AAS History Series*, pp. 137-201.

Chapter 8

Reaction Motors, Inc.: A Corporate History, 1941-1958[1]

Part II, Research and Development Efforts

Frank H. Winter[2] and Frederick I. Ordway, III[3]

The idea of employing a rocket as an auxiliary powerplant for aircraft hardly originated with RMI. Perhaps the earliest concept for Jet-Assisted-Take-Off (JATO) was conceived by T. J. Bennett of Oxford, England, in the 1880s. Bennett made a 30 lb (13.6 kg) model of a steam-propelled aircraft but quickly realized that his steam engine was underpowered for a rapid take-off from a short surface. His solution was to provide an initial boost to the model by means of gunpowder rockets. "The starting was very successful," he recalled years later, "I wish I could say the same of the subsequent flight."[4]

Jato

H. H. Bales of Ashcroft, British Columbia, Canada, took out probably the first JATO patent in which he, too, incorporated skyrockets attached to an aircraft. This was

---

[1] Presented at the Sixteenth History Symposium of the International Academy of Astronautics, Paris, France, 1982.

[2] Department of Space History, National Air and Space Museum, Smithsonian Institution, Washington, D.C., U.S.A.

[3] The Space and Rocket Center, Huntsville, Alabama, U.S.A.; formerly of the Engineering Division, Reaction Motors, Inc.

[4] Letter of T. J. Bennett, *Flight*, 26 March 1910, p. 244.

U.S. patent No. 1,003,411, 19 September 1911, for a "Pyrotechnical Auxiliary Propelling Mechanism." Other patents followed, such as one by the French aviator, Albert Lepinte, who envisaged extra power from rockets more as a safety feature; rockets could be reversed so that they retarded an aircraft's rapid descent or accelerated the plane in time of danger.[5] The idea was certainly in the air at this time, because a few years later, in the Soviet Union, V. I. Dudakov and V. A. Konstantinov took out their own patent on 8 October 1928 for the "Construction of Rocket Engines for Powered Aircraft Flight," which in essence was also a JATO.

These gentlemen may or may not have learned of probably the first full-scale successful JATO flight made on 8 August the following year, when a Bremen Type Junkers W33 land plane fitted with floats took off from the Elbe River near Dessau, Germany, using six powder rockets ignited rapidly in pairs. Although the Junkers firm was convinced it would one day be possible to use rocket assist to lift the then large load of 11,000 lb (5,000 kg), follow-up experiments did not come until much later. In any case, Dudakov and Konstantinov continued their own researches for the Soviet Army-supported Gas Dynamics Laboratory (GDL). Their solid-propellant rocket assisted flights began in 1931, first with a U-1 light aircraft, then afterwards with TB-1 bombers. They later progressed to liquid-propellant JATOs, but these developments were totally unknown outside the USSR and have only recently come to light.

Most likely, the highly sensational rocket-assisted glider stunts of 1928-1929 by the colorful German automobile manufacturer, Fritz von Opel, precipitated serious JATO researches. Besides the Soviet Union, the military services of Germany, the United States, Great Britain, Japan, and Poland undertook JATO development in earnest by the mid-1930s, albeit some of these programs were short-lived.

The earliest American JATO work is of particular interest in the RMI story. The U.S. Army was generally apathetic towards rocket weapons in the 1930s, but the Air Corps did support the famous Guggenheim Aeronautical Laboratory of the California Institute of Technology (GALCIT) and its efforts to develop practicable rockets from 1938, including JATO. The GALCIT work eventually led to the formation of the Aerojet Engineering Corporation in 1942 and to establishment of Caltech's Jet Propulsion Laboratory (JPL) in 1944; the latter conducted rocket research, the former initially conducted JATO manufacture and development.[6]

Whereas Aerojet's principal customer for this product was the U.S. Army Air Corps, the first and a long-time customer for RMI was the U.S. Navy. RMI began shortly after the bombing of Pearl Harbor by the Japanese in 1941, and even earlier (in

[5] British specification No. 229, 670 of 19 February 1924 with a later U.S. patent, No. 1,611, 353 of 21 December 1926. A dramatic drawing of Lepinte's deceleration rockets in action may be found in the French magazine *Je Sais Tout*, 24 Annee, June 1928, p. 170. See U.S. Patent 1,003,411 of 19 September 1911 to H. H. Bales; British Patent 229, 670 of 19 February 1924 to A. Lepinte; and V. I. Dudakov, "Analysis of the Early Period of Using JATO in Aviation" (in Russian), *Proceedings of the 13th International Congress of the History of Science*, Moscow, 18-24 August 1971, pp. 44-46.

[6] Aerojet, the U.S.'s second pioneering commercial rocket corporation, was similarly interested in multiple applications of rocket power, but responded to the immediate wartime need for JATOs on heavily-laden military aircraft.

November 1941) RMI's future President, Lowell Lawrence, succeeded in arranging for a test run of James H. Wyld's regeneratively cooled motor for a Navy representative.

Yet the Navy soon came to support not one but five liquid-propellant JATO projects simultaneously. These were: RMI; work at Aerojet and GALCIT; a Navy group headed by Robert H. Goddard; and the Navy's own research team headed by Lt. Comdr. Robert C. Truax. Truax, who like Wyld had his original rocket engines tested on the American Rocket Society's proving stand No. 2 in 1938, began his own Navy JATO work late in 1941. Goddard signed his joint Navy and Army Air Corps contracts at about the same time, but moved his personnel and equipment to the Naval Engineering Experiment Station at Annapolis, Maryland, in July 1942. In fact, both RMI and Truax conducted their JATO projects at the Annapolis station.[7]

There were several reasons for this apparent duplication of effort. According to Lou Arata, one of the first shop workers hired by RMI, rocketry was then a wide-open field with so many variables that needed testing out that "even more groups or experimenters would have been welcome." Truax, a Navy man and ardent rocket experimenter himself, confirmed this view and added that there was also tremendous pressure at the time because of the Pacific war, so that the Navy was very anxious to "try as many different parameters and designs as possible for the greatest chance of success within the shortest time." As for stationing three of the five JATO groups at Annapolis, the reasons here were threefold: the facilities of the Naval Engineering Experiment Station (NEES) had existed for decades and could be put to good use on the JATO projects; the VN-8 Naval Seaplane Squadron for testing the JATOs was conveniently right next door along the Severn River; and Annapolis was close to Washington, where Navy JATO projects were administered.[8]

The three Annapolis groups indeed offered different approaches and were always friendly and even helpful to the others. Goddard, for example, noted in his diary for 15 September 1942: "Went with Lt. [C. Fink] and Mrs. Fischer to New Jersey, and saw the Reaction Motors setup. Had dinner with them and Lawrence and Shesta."[9] Arata also remembers occasional visits from Goddard and Truax ,and that the former loaned a pump to RMI. Unfortunately, it was defective and leaked! Nonetheless, Goddard's papers reflect a continued close and sincere association with both RMI and Truax throughout Goddard's Annapolis years until his death in 1945. On one or two occasions, Lawrence suggested a merger with Goddard, but without results.[10]

The chief difference in approach to Navy JATOs was the choice of propellants. Goddard consistently favored liquid oxygen (LOX) and gasoline since he started with liquids in 1920. Truax, who had independently developed his own regeneratively-cooled

[7] See "JATO" file, National Air and Space Museum, Smithsonian Institution, Washington, D.C. RMI was assigned a small building at this site for part of their work in late November 1943.

[8] Interview with Lou Arata by Winter, 9 June 1982; interview with George S. James by Winter, 8 June 1982; and interview with Robert C. Truax by Winter, 7 June 1982.

[9] C. Fink Fischer was RMI's liaison officer with Navy.

[10] Esther C. Goddard and G. Edward Pendray, eds., *The Papers of Robert H. Goddard* (New York: McGraw-Hill, 1970), 3:1478, pp. 1558-1560, 1572-1573.

rocket engine in the late 1930s, began with compressed air and gasoline. When he switched to Goddard's combination, he encountered ignition and combustion problems. Then, in 1942, a solution was found: one of his assistants, Ray C. Stiff, made the serendipitous discovery of the combination of nitric acid and aniline, which are hypergolic (spontaneous ignition). Artificial means of ignition were not needed and combustion was smooth. Another advantage was that super cold liquid oxygen and associated cryogenic equipment did not have to be used—a boon for wartime military operations. In modern parlance, the propellants were "storable" or "packaged" in that they could be kept at room temperature. Greater specific impulse was also gained. On the negative side, the acid was extremely corrosive and did require special handling. Truax still shifted to the hypergolic, a move followed by Caltech and Aerojet; but Goddard stubbornly clung to the standard LOX and gasoline, always objecting to the acid's corrosiveness and toxicity.

To establish priorities, nitric acid as an oxidizer instead of LOX is alleged to have been tried as early as 1930 by Friedrich Sander in Germany, though he was more well known for his solid-propellant (gunpowder) motors. In 1932-1933, the American, Harry Bull, tested nitric acid and turpentine, a mixture found to be not particularly smooth. The Germans secretly began to develop their own hypergolic mixtures from the late 1930s; Helmut von Zbrowski was the most prominent among the experimenters. These efforts culminated in projects such as the Wasserfall and Taifun missiles of the mid-World War II period. At RMI, this form of propellant was not fully investigated until the mid-1940s, when it was employed in the Navy's Gorgon II-A and Lark programs.[11]

RMI's original Navy contract for JATOs, Project TED-EES-3401, in early 1942, stipulated that the company deliver the existing Wyld regeneratively cooled liquid oxygen-alcohol motor. In addition, an identical, though 100 lb (45.4 kg) thrust liquid oxygen-gasoline version was also to be demonstrated. The third and most important part of the contract called for the development of a 1,000 lb (454 kg) thrust LOX-gasoline unit. This engine had to demonstrate repeated starts and was to be completed within 180 days.

In their own interests, the RMI founders felt it best not to deviate from the LOX-gasoline staple, although the Wyld engine was designed only for the alcohol fuel. According to Shesta, "A replica of the original Wyld motor quickly burned out when tested with gasoline. This would never do. A complete re-design was necessary." Among the changes was a substitution of a copper nozzle for the former one machined from aluminum. This new motor was designated M15-G-1. RMI called it the Wyld Regeneratively-Cooled Rocket Motor Serial No. 2.

From this simple unit evolved several other small experimental motors, all of which were tested by RMI personnel on the American Rocket Society's test stand No. 2, which is also on display in the Museum. Since the stand had largely been designed by RMI founder John Shesta, and had been built in co-founder Franklin Pierce's basement, it was easily "borrowed" by the fledgling company—especially after the Society ceased

[11] Frank J. Malina, "The U.S. Army Air Corps Jet Propulsion Research Project, GALCIT Project No. 1, 1939-1946: A Memoir" in R. Cargill Hall, ed., *Essays on the History Symposia of the International Academy of Astronautics* (Washington, D.C.: (NASA Conference Publication 2014, 1977), 2:167-168; reprinted in Vol. 7, Part II, *AAS History Series*, (San Diego: AAS Publications Office, 1986), pp. 167-168.

its experimental activities following the bombing of Pearl Harbor. This stand could statistically fire rockets up to 200 lb (90 kg) of thrust.[12]

Eventually, RMI was compelled to build a larger stand. This was a permanent stand (ARS stand No. 2 was portable) anchored to a concrete slab at Franklin Lakes, New Jersey, and had an initial capacity of over 1,000 lb (454 kg) thrust. An adjacent blockhouse, which still exists and was recently named an official historic landmark, was constructed 12 ft. (3.7 m) away. A surviving handwritten RMI logbook indicates that Wyld had much to do with designing these testing facilities. One interesting feature was a movable security shelter that was rolled over the stand when it was not in use, although the old German Rocket Society (VIR) erected a similar construction about 1933.

The Wyld serial No. 2 and similar small motors proved that gasoline and LOX could adequately work in a regenerative system. The final motor of this series was run about 50 times for durations of a few seconds to more than a minute. Even as this work was conducted, progress was made toward the development of the 1,000 lb (454 kg) thrust motor. Because of its larger size, greater operating temperatures were anticipated. Consequently, stainless steel was chosen for the nozzle rather than copper. Drawing from his past experience with this material, Shesta designed a set of punches and dies. With these, and a hydraulic press located by Lawrence, "very satisfactory nozzles" were formed.

Lawrence was also responsible for developing remotely-controlled throttling valves operated by small electric motors and gears. Wyld and Lawrence applied for US Patent No. 2,479,888, on 6 July 1943, for their throttling system; a patent titled "Controlling System for Reaction Motors" was not granted until 23 August 1949.[13]

Spark plug or squib igniters often proved troublesome. In one test, the spark plug porcelain broke, causing a spray of oxygen to escape from the hole. Ignition delays, improper pre-mixing of propellant, and lack of ignition frequently occurred. Despite these problems, the 1,000 (464 kg) thrust unit was satisfactorily delivered to the Navy on schedule; although there is one RMI legend that the original engine blew up shortly before the qualification test was to be held and a replacement was built in record time.

By the time the 1,000 lb (464 kg) unit was finished, RMI had produced at least ten different models of motors (not including test blanks), ranging from 50 lb to 1,000 lb (23 to 464 kg). However, work was already proceeding with a 2,000 lb (900 kg) unit. This too, was delivered to the Navy, which was now convinced that much larger models were feasible. A second contract called for a 3,000 lb (1,360 kg) motor that was meant for installation in a Martin PBM patrol bomber.[14]

---

[12] Interviews with John Shesta by Winter, 10 June 1982, and by Ordway, 3-6 July 1982.

[13] Reaction Motors, Inc., handwritten (pencil) logbook by James H. Wyld, *et al.*, 1943-1945, donated to the National Air and Space Museum by John Shesta; Shesta interviews; and U.S. Patent 2,479,888 of 23 August 1949 to James Wyld and Lovell Lawrence.

[14] RMI Logbook and Arata interview.

Shesta made an interesting innovation in this new engine. He fitted a multi-hole (40 or more openings) oxygen injector into the injector plate. These holes were aligned concentrically with holes in the lower plate that fed the gasoline. "These holes," Shesta reported, "protruded slightly past the lower plate, the annular space providing a clearance for a thin film of gasoline feed. As the oxygen emerged from the central hole, it spread out in a cone that intersected the annulus of fuel and provided good mixing."

Arata recalled that informality marked the early days of RMI, with the modifications of designs and techniques such as those worked out over lunches or coffee breaks at the nearby Triangle Grill. "Usually," he remarked somewhat facetiously,

> we met there to discuss an explosion that had occurred the day or morning before. We would all pool our heads together to figure out what went wrong, where the obstruction or the hot spot was. Then one of us would say, "Why not try such-and-such?' "Bun," a nickname for Lawrence, would respond, 'Yeah, let's try it!' Then we would all get back and work like hell, sometimes until 9, 10, or 11 at night. Sometimes John Shesta would also come around with a clipboard and just make up modifications on-the-spot.

"Our greatest problem during those years," Arata continued, "was heat transfer. Of course, regeneratively cooled systems were the only way to go; but this did not mean that all the cooling problems were solved, especially for long-duration runs."[15]

One answer was Pierce's idea of making a corrugated, helical winding or spiral out of a metal ribbon. The windings were placed between the cooling jacket and liner for swirling the fuel. The object was to slow down the circulation process, so that the motor would be better cooled. The fuel was preheated for better combustion.

The difficulty with this technique was that the ribbons were not always placed in the right positions. After some faulty test runs, the motors were dismantled to learn the difficulty. Often it was found that the ribbons were jammed in such a way as to cause fuel blockage rather than to assist the flow. Sometimes, unknown to the RMI workmen, the ribbons shifted during the manufacture, prompting the staff to dub them derogatorily, "wiggle rings." After one or two Triangle Grill sessions, this problem, too, was overcome. The rings were replaced with wires that fitted into spiral grooves around the cooling liner. By adjusting them at different angles and lengths, it was possible to regulate the coolant flow.[16]

The completed 3,000 lb (1360 kg) thrust motor weighed 75 lbs (34 kg) with auxiliaries (nitrogen pressure bottle and other tanks). The chamber was about 6 in (15 cm) in diameter and 6 ft (1.8 m) long. A run conducted on 6 May 1943, according to Wyld's handwritten note in the RMI logbook cited above, lasted 24 seconds at high thrust of 3,180 lb (1442 kg) maximum. "Flame straight," Wyld observed, "not remarkably long but intensely hot—started a number of small fires in woods and scorched ground. Also burned corners off some nearby sandbags . . . Motor appears to be in excellent shape after run . . ." (Figure 1).

---

[15] Arata interview and Shesta interviews.

[16] Arata interview.

In subsequent tests, igniter problems cropped up; but on 14 May 1943, a "very long run, probably over 1 min" was recorded. This became the standard duration of the engine.[17] By October, the first static tests were made with simulated PBM-3 aircraft boattails made of plywood and covered with corrugated metal. This phase of the testing was undertaken in RMI's New Jersey plant, (Figure 2).

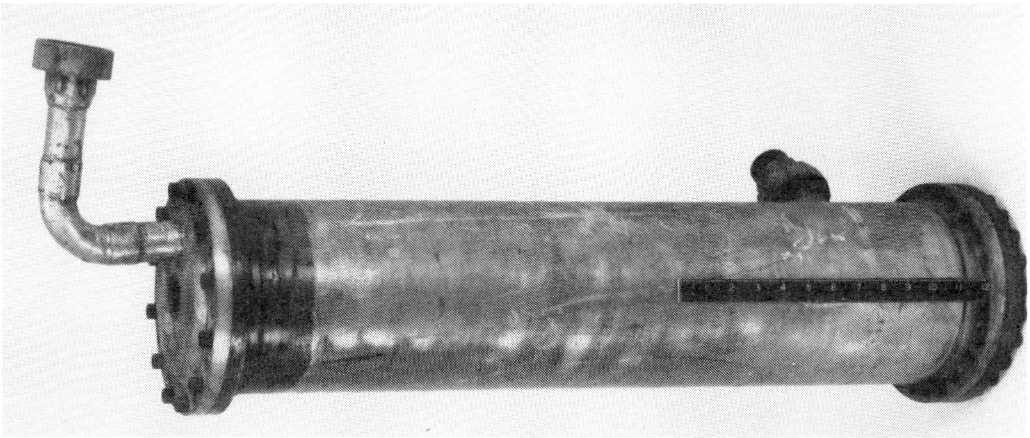

Figure 1 Reaction Motors, Inc. 3000-A-1 rocket engine, a 3000-lb. thrust JATO (Jet-Assisted-Take-Off) unit for heavily loaded U.S. Navy seaplanes, developed 1942-1945. Courtesy of Frederick I. Ordway III.

Figure 2 Reaction Motors, Inc. 3000-A-1 JATO (Jet-Assisted-Take-Off) rocket on a Martin PBY aircraft tail section mockup at Reaction Motors plant, Pompton Plains, New Jersey, ca. 1943. Courtesy of Frederick I. Ordway III.

[17] RMI Logbook.

Finally, the time arrived for the real tests on aircraft taking off from water. Captain William L. Gore, US Marine Corps, served as the pilot of the first RMI "live" JATO flight at Annapolis on 12 January 1944. (He was already a veteran of similar flights since he managed Goddard's JATO on a PBY in September 1942.) By all accounts, the RMI trials went well. Yet for all the RMI labors, the Navy never officially adopted any of the company's JATO units.

What had gone wrong? In a word, it was RMI's choice of liquid oxygen and gasoline. Compared with the proven efficiency and convenience of hypergolics used by Truax and Aerojet, LOX-gasoline was not considered a practical alternative. The Navy was especially concerned about emergency wartime use of JATOs, in which storable (non cryogenic) propellants had the edge. In the final analysis, Aerojet's castable solid-propellant JATOs were the most successful of all the JATO projects because they combined easy storage with simplicity and economy.

At the conclusion of the PBM tests, Shesta reminisced that "we had definitely given up gasoline as a propellant and switched to alcohol, which we felt was much better suited for rockets." This higher performance and cooler fuel was subsequently used in the RMI's other rocket projects.[18]

Power for Aircraft—Rocket Planes

Ironically, while the hypergolic combination (red-fuming) nitric acid and aniline possessed desirable qualities for JATO applications, it had very serious drawbacks that made it undesirable for aircraft propulsion. Rocket plane pilots would be exposed to great risks. Nonetheless, these weaknesses were felt to be soluble in time, and the non-igniter advantage of hypergolics led Aerojet to develop, by 1944, a 2,000 lb (900 kg) thrust, acid-aniline engine for the proposed Northrop XP-79 rocket-propelled, flying-wing interceptor for the Army Air Corps. This unit was called the "Aerotojet."[19]

The plane did not materialize, and the Aerotojet was not fully developed; but by late 1944 two transonic aircraft projects began to emerge. One was the confidential Navy proposal of a rocket-powered interceptor, possibly the adaption of an auxiliary rocket powerplant on an existing Navy plane. The other project, sponsored by the Army Air Corps, was an air-launched, all rocket powered aircraft known first as the Bell XS-1 (Experimental Supersonic-1), and later called simply the X-1.[20]

In January 1945, the Navy's Ships Installation Branch was requested to develop a rocket powerplant using LOX and gasoline for the secret interceptor, designated D-558. "At that time," according to the 1948 GALCIT survey *Jet Propulsion*:

---

[18] C. W. Schnare, "Development of JATO and Engines for Manned Rocket Aircraft," *American Rocket Society Paper 2088-61*; and Shesta interviews.

[19] James interview.

[20] Richard P. Hallion, *Supersonic Flight* (New York: Macmillan Co., 1972), pp. 20-21 and 39-40.

no data was available for specific powerplant requirements. Accordingly, Reaction Motors, Inc. of Pompton Plains, N.J., was requested to undertake the development program on a small size powerplant. This approach was taken in order to anticipate and solve problems as soon as possible in small scale and hence save time as well as expenditures of money and equipment associated with large scale tests. Numerous problems in motor ignition and gas feed control were worked on in this manner during February, March and April of 1945. In the early part of May it was decided that the power plant should deliver between 4,500 pounds 2,040 kg and 6,000 pounds 2,720 kg thrust. Consequently, the total thrust was divided into four parts of 1,500 pounds 680 kg thrust each. By this method, the fourth motor can be added or deleted as required and the composite powerplant can be operated at any one of the immediate stages.[21]

This four-chambered motor became famous as the 1500N-4C and later the 6000C4, referring to its 6,000 lb (2700 kg) thrust (Figure 3). It operated on LOX and alcohol/water rather than LOX-gasoline.

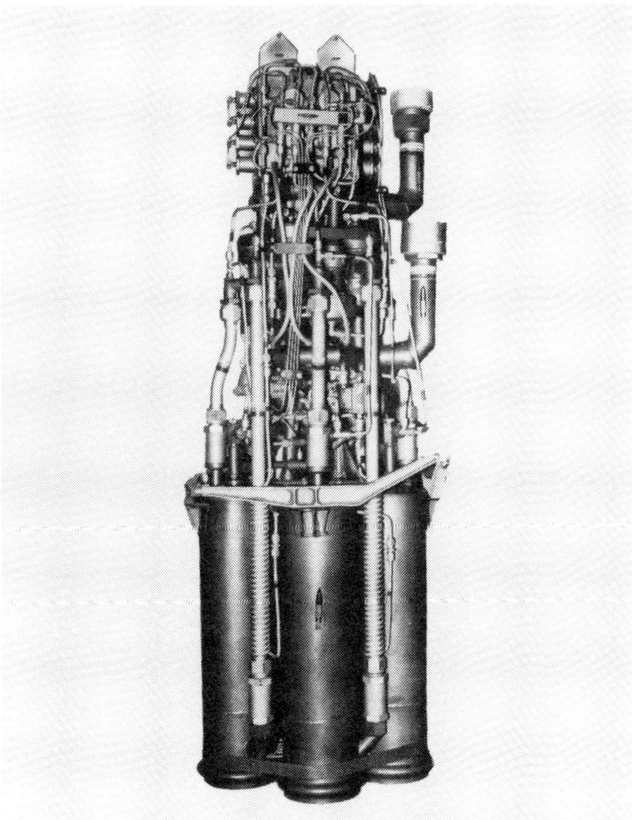

Figure 3 Reaction Motors, Inc. 6000C-4 rocket engine. Courtesy of Frederick I. Ordway III.

[21] Guggenheim Aeronautical Laboratory, California Institute of Technology (GALCIT) Survey, *Jet Propulsion* (Pasadena: California Institute of Technology, 1946), pp. 136-137.

"In cooperation with the Army Air Force's" program of transonic aircraft research," the GALCIT survey continued, "the Navy has arranged to release the first propulsive unit of the above design to the Bell Aircraft Corp." This is how the 6000C4 started under a Navy contract and yet first powered the Air Force-sponsored Bell X-1. As for the fate of the confidential rocket-propelled interceptor, this is unknown. But evidently the project never materialized.

According to Harry W. Burdett, a key RMI engineer on the 6000C4, the original engine's outer jacket was made of mild, or carbon steel (Shesta says it was chosen for reasons of economy). Carbon steel rusts; the engine also had copper tubing. Hence, when Bell Aircraft's Robert M. Stanley visited RMI, the engine was somewhat rusted and did not look presentable. Shesta remarked: "I don't like that. It's got to be stainless steel." The change was made and the engine painted black. Burdett says that because of this appearance, plus the tremendous noise it made when fired, the 6000C-4 was affectionately called the "Belching Black Bastard." However, he adds, this name had to be "cleaned up for the press." Consequently, it was named "Black Betsy," after Lawrence's infant daughter, Betsy, a name that also signified reliability. Officially, the Navy designation was LR-8. The Air Force designated it the LR-11 (Liquid Rocket 11).[22]

The first 6000C4 was fired on 31 August 1945. All the individual cylinders were fired separately, but no effort was made to fire them simultaneously, since the oxygen manifold pressure dropped when the third cylinder was ignited. It is not known when the first simultaneous firing of all cylinders was made, but it is known that from late August, the engine, or separate cylinders, were fired almost daily. Blowouts, ignition and injection problems were common in the development of the engine. But Lou Arata, RMI's self-taught expert shop man, adapted his own igniter, which was successfully tested 500 times without failure. It was the miniature rocket similar to the scheme mentioned above. The igniter was shaped like a venturi with fuel and oxidizer inlet holes drilled in the sides. The fuel and oxidizer were premixed and ignited by a spark plug.

Lawrence, who at that time had just returned from Germany after participating in the US Government's post-war investigations of Germany's missile technology, was so impressed with the igniter that he asked Arata to immediately make four of them. The igniter, according to Arata, was the basis for the one subsequently adopted for the 6000C4 by Harry Burdett, the project engineer.

Prior to passing its qualification test, the 6000C4 was subjected to more than 800 seconds total firing on RMI's specially designed test stand, which simulated the roll, pitch and yaw of an aircraft. This was a large boom on which the engine could be mounted and fired in horizontal and vertical positions. Contracts C-107193 and C-108792 were awarded to Bell Aircraft, effective 29 May 1946, with a scheduled completion date of 18 April 1947 for delivery of the 6000C4, which was qualified under RMI Project No. 171.[23]

---

22 Interview with Harry W. Burdett, Jr., by Winter, 25 June 1982.

23 Burdett interview; Arata interview; Reaction Motors, Inc., untitled, unpublished list of RMI projects and project numbers with dates, 1946-1955, in "Reaction Motors, Inc." file, National Air and Space Museum.

The subsequent history of RMI's 6000C4-powered aircraft, Bell XS-1 (X-1), also popularly called "Glamorous Glennis" after the pilot's wife, is well covered in the literature. On 14 October 1947, Captain Charles "Chuck" Yeager flew it and became the first man to break the sound barrier. He flew at Mach 1.06 or 700 mph (1,130 km/hr) (Figure 4).

Figure 4 Bell X-1 rocket airplane making the world's first supersonic flight, 14 October 1947, Captain Charles E. Yeager, pilot. The aircraft was powered with a 6,000-lb thrust Reaction Motors, Inc. A6000C-4 engine. Courtesy of Frederick I. Ordway III.

Other data relative to the 6000C4, which made this great advance possible, are that it burned an ethyl alcohol and water mixture with LOX. The engine was 54 in (137 cm) long and 49 in (124 cm) in diameter. (The arrangement of the four cylindrical thrust-chambers for the Air Force's X-1 was a diamond pattern; for the Navy's later D-558-2 and other projects a square pattern was preferred.) Each chamber was 18 in (45 cm) in outside diameter. Total engine weight for the Bell X-1 version (nitrogen-fed) was 345 lb (156 kg) empty. Later 6000C4s used a feed system employing an RMI-developed turbopump driven by the decomposition of hydrogen peroxide at a speed of 12.2 rpm.[24]

[24] Reaction Motors, Inc., Liquid Propellant Rocket Engines Developed for the U.S. Military Services. Formerly a Confidential Handbook, Looseleaf, ca. 1952.

The 6000C4 also powered the Bell X-1A and Bell X-1-3 aircraft. With slight modifications, it was fitted in the Navy's Douglas D-558-1 Skystreak, which could also be powered with an Allison J-35-A-11 5,000 lb (2,270 kg) static thrust turbojet. Similarly, Black Betsy served as the powerplant for the all-rocket Douglas D-558-2 Skyrocket and the Douglas D-558-2 Skyrocket, jet and rocket version. Some of the D-558 series were modified for air-launches. These aircraft also flew supersonic. One of the planes, the D-558-2 No. 2, jet and rocket version, became the world's first aircraft to exceed twice the speed of sound, reaching Mach 2.005 or 1291 mph (2,077 km/hr) during a dive on 20 November 1953; Scott Crossfield was the pilot.[25]

The first rocket aircraft were experimental and designed to study transonic flight. The Air Force's Republic XF-91, fitted with a variation of the 6000C4, was the first supersonic combat fighter; though technically it was experimental. Originally, the XF-91 was fitted with a 16,000 lb (7,260 kg) Curtiss-Wright XLR-27-CW-1 rocket engine but there were serious technical problems with it, and Republic then switched to the lower-powered but reliable 6000C4. Like the Skyrocket, the XF-91 had dual powerplants. The rocket was considered an auxiliary means of propulsion for accelerated take-offs and climbs and for operations at altitudes beyond the sonic barrier. The XF-91's General Electric J-47 turbojet of 5,200 lb (2,360 kg) of static thrust provided primary power. In the XF-91, the 6000C4 engine was placed below the turbojet exhaust exit.[26]

In September 1952, Republic's test pilot, Russell (Rusty) Roth, flew the XF-91 on its first supersonic flight. The company's president, Mundy I. Peale, announced the flight shortly afterwards, stressing that the XF-91 was "not purely a research plane" and "is combat ready." Only two of the aircraft were built, however, and no production contract was let.

This was not the end of RMI's aircraft rocket powerplants. On 30 September 1955 a contract was let to North American Aviation Corporation for the development of the hypersonic X-15, a major step toward the Space Shuttle. Because of its experience and success with the earlier X-series of aircraft, RMI was awarded the contract for the X-15's engine, which became the fully throttable 59,000 lb (26,760 kg), maximum-thrust XLR99 Pioneer rocket that used anhydrous ammonia and liquid oxygen. The first flights made with the X-15, from 1959 to 1960, were accomplished (often past Mach 2) with two modified Black Betsy engines with the Air Force designation of XLR-11.[27]

Test and Sounding Rockets

Following a close second in importance to its aircraft auxiliary and principal rocket motors was a series of propulsion units developed for Navy and Air Force missile

---

[25] Hallion, *Test Pilots*, pp. 203-205, pp. 172-173.

[26] "Rocket-Powered Fighter Breaks the Sound Barrier," *New York Times*, 10 December 1952; and Albert M. Skea, "New Fighter Plane Fast," *Evening News* (Newark, New Jersey), 10 December 1952.

[27] *Ibid.*; Eugene M. Emme, ed., *Aeronautics and Astronautics 1915-1960* (Washington, D.C.: Government Printing Office, 1961), p. 79.

and atmospheric sounding vehicles. Included were rocket motors for the Lark surface-to-air test missile, the MX-774 precursor to the Atlas intercontinental ballistic missile and the Viking rocket used at first for extreme altitude exploration and later for testing components in the Vanguard satellite launch vehicle.

Lark LR2-RM-12 and XLR6-4 Engines

During 1944 and 1945, RMI built approximately fifty 350 lb (160 kg) thrust rocket engines that operated on mixed acid and mixed aniline hypergolic (self-igniting) propellants. Based on engineering drawings supplied by the US Naval Engineering Experiment Station of an earlier 340 lb (154 kg) CML-2N-1 design, they were mounted on the Gorgon II-A pilotless aircraft. The Navy planned on using these missiles against Japanese naval and merchant shipping during World War II.

The credit for this development is due to the Navy's Lt. William Schubert and Ensign Robertson Youngquist. According to John Shesta, "the Gorgon-Lark units were constructed by us [RMI] in accordance with their specifications. All we did was to refine some manufacturing techniques and nozzle construction." Before delivery to the Navy, each unit had first to be acceptance-tested.

The nitric acid-aniline combination was subject to ignition delay, resulting in hard starts. The problem, according to Shesta, was solved "in an elegant way" by Youngquist's pintle valve device. "The evil smelling products of combustion," Shesta recalls, "did nothing to endear us to the local population."

Building on CML-2N-1 experience, Al Huse, Edward Francisco, Ray Pepplar and other RMI engineers developed a two-chamber unit consisting of 220 lb and 400 lb (100 kg and 181 kg) thrust chambers mounted together (the latter was an uprated Gorgon 11-A cylinder). Two models evolved, the pressure-fed LR2-RM-12 and the turbopump-fed XLR6 RM-4. About 500 of these engines were built for the Lark ground and ship-to-air experimental missile. The concept was to use the smaller cylinder for cruising and the larger one for extra velocity as the target was approached, see Figure 5.

Figure 5 220-lb thrust rocket chamber of the two chamber CML-5N rocket engine with a total thrust of 620 lbs. Produced for the U.S. Navy "Lark" missile by Reaction Motors, Inc., 1944-1945. Chamber in the National Air and Space Museum. Courtesy of Frederick I. Ordway III.

Manufactured both by the Fairchild Engine and Airframe Company and the Consolidated Vultee Aircraft Corporation, Lark was originally designated as the XSAM-N2a. Later, in 1950, it was reclassified as the CTV-N-9 component test vehicle. Most of these were fitted with Mach limiters that permitted the missile to reach, but not to exceed, Match 0.85. When a Lark approached this velocity, an indicator closed the valve, which, in turn, closed the propellant valve and consequently stopped the operation of the 400lb (181 kg) thrust chamber, The reverse occurred when the missile dropped below the limiting velocity. Launchings at first took place from a 450 ft (140 m) long inclined ramp at the Naval Ordnance Test Station, Inyokern, in California. Later, they were made from a zero-length launcher at the Naval Air Missile Test Center in Point Mugu, also in California, as well as from the USS Norton Sound at sea. Typically Lark would undergo 240 seconds of powered flight; but, in a test in December 1948, power lasted for 295 seconds, carrying the missile over a distance of more than 25 mi (40 km). Although the 620 lb (281 kg) thrust engines were employed exclusively for Lark, RMI felt that they could be used as assist take-off units for light liaison and command-type aircraft.[28]

MX-774's SLR35-RM-1 Engine

After the war, RMI carved out a small piece of the emerging large rocket-propelled missile programs that were based on, or inspired by, the V2 and other German military vehicles. During the immediate postwar period, a German team under Wernher von Braun and its American colleagues began flying and testing captured V2s from the White Sands Proving Ground in New Mexico, while simultaneously embarking on the new Hermes missile project. For its part, the Navy had started developing a Viking sounding rocket that was based in considerable part on V2 technology. At the same time, the newly created Air Force was tackling a long-range rocket-powered missile using the MX-774 to test the concept.

As it turned out, RMI received engine development contracts for both the Air Force and Navy projects. Though short-lived, the MX-774 did provide the Air Force with its airframe contractor, the Consolidated Vultee Aircraft Corporation, and RMI with valuable experience. Its motor, designated the XLR-35-RM-1, was an advanced version of the four-engine, regeneratively cooled 6000C4 engine, which had already been successfully flown. Feed pressure was increased to about 425 psi (30 kg/cm^2), commercial 95 per cent ethanol served as its fuel, and liquid oxygen the oxidizer. The motor operated over a nominal 7,600 lb to 8,400 lb (3,450 to 3,800 kg) thrust range.

The mounting of the four cylinders was arranged in such a way that they could swing back and forth. Propellants were fed through two coaxial seals, one on each side of the motor, which also served as bearings. Movement was controlled by a system of hydraulic cylinders. Many difficult design problems were worked out by an RMI engineering team led by Peter Palen. Since the XLR35-RM-1 was to power an unmanned

[28] Shesta interviews; and Frederick I. Ordway III and Ronald C. Wakeford, *International Missile and Spacecraft Guide* (New York: McGraw-Hill, 1960), pp. 187-188.

missile, it needed to be ignited only once. In the light of this, RMI engineers were able to delete the 6000C4s igniter assembly.

Extensive work was also done on the LOX face seals to permit the rotational movement of the thrust chambers on the mounting structure as well as the LOX propellant valve located in the middle of the same structure. This was undertaken by M. E. (Bud) Parker and Richard Weiss with the assistance of Frank Coss.

Shesta emphasized the importance of assuring that all the four cylinders would ignite simultaneously. If any cylinder failed to ignite, while the others had successfully fired, he warned "a dangerous situation would ensue."

The ignition problem for vertically fired rockets was of special concern to him. He recalled that in the early American Rocket Society days, someone would be elected, or perhaps even volunteer, to ignite a rocket. "He would dash up . . . with a gasoline rag on the end of a stick, light up and run like hell. Occasionally he would be wearing an asbestos suit."

When Shesta joined the Society he refused to get involved in what he termed "foolishness." "We began mixing our own powder," he continued, "and making squib igniters. These were started electrically." Later, at RMI, Lawrence obtained a supply of meal-powder from Dupont's. This worked fine, but the combustion was too rash. There was not enough time to make sure of reliable ignition. The timing was too critical. One time I asked Jimmy Wyld to mix up a batch of powder with other things to make it burn slower and safer. He did, and he succeeded beyond our fondest hopes. It was the safest powder in the world. You could ignite it with an acetylene torch, but as soon as you pulled the torch away it would go out."

For the MX-774 rocket, Shesta asked Dr. David Horvitz of the RMI research laboratory to come up with a good igniter. After a few experiments, he developed a composition of potassium perchlorate combined with a plastic resin which, after being baked in the oven, resulted in hard smooth rods about 0.25 in. (0.63 cm) in diameter and some 4 in. (10 cm) to 6 in. (15 cm) long. "These," recalled Shesta, "would burn with a fierce hot flame for about 30 seconds. In fact, they were perfect igniters." Shesta explains how they were used:

> One igniter cartridge was inserted in the nozzle of each cylinder. The ignition was by means of a hot (tungsten) wire 0.005 inches in diameter—the four were connected in series. A short distance beyond the ignition point, the rod was drilled and a copper wire was passed through the hole. When the fire reached the copper wire, it would melt it. These wires were connected in parallel. When all the four wires melted, they would actuate a relay, which in turn would turn on the propellants. If only three wires melted and one remained intact, nothing would happen.

The engine weighed 396 lbs (180 kg). The decomposition products of hydrogen peroxide drove the turbine, which was coupled to two pumps that furnished the propellants to the combustion chambers.

The design of XLR35-RM-1 was completed in late 1946, and model and acceptance testing was undertaken in June of the following year. A maximum thrust of 8,800 lb (4,000 kg) was obtained at a specific impulse of 227 seconds. James Fitzgerald and

Frank Coss of RMI took the unit through to integrated testing and then to the White Sands Proving Ground for flight testing.

The MX-774's airframe and guidance systems were designed by a Convair team headed by Karel J. Bossart, and it consisted of three metal sections; the fuselage nose, fuselage center and the tail. The nose section was made of aluminum alloy and housed the electrical, stabilization, and guidance and control components, and test instrumentation. The center section consisted entirely of propellant tankage constructed of welded aluminum. The liquid rocket engine and the controlled surfaces were housed in the aluminum alloy structure in the tail section. Integral tanks were introduced to rocketry with the MX-774, separating the nosecone from the main rocket structure after peak altitude was introduced.

The MX-774 was static-tested at Point Loma, San Diego, in California, from mid-1947 to late 1948. It was flown three times at White Sands in July, September and December. A failure in the electrical system led to premature cutoff of an otherwise flawlessly functioning first flight engine. MX-774 rose over 1 mile (1.6 km) high, then crashed only 600 ft (180 m) from the blockhouse. No. 2 fared somewhat better, reaching nearly 30 miles (48 km). Excessive oxygen tank pressure caused the missile to break apart in mid-air. The third flight went well for 51 seconds, when the engine cut off. An attempt was made to parachute-recover the missile, but vibrations caused it to explode. The MX-774 project is credited with demonstrating the well-known practice of steering by vectoring the rocket motors.

The Viking XLR10-RM-2 Engine

The engine developed for the Naval Research Laboratory's RTV-N-12 Viking high-altitude sounding rocket was quite different from that selected for the MX-774. Instead of four small thrust chambers it relied on a single, large one that produced more than 20,000 lb (9100 kg) of thrust (Figure 6).

The contract to develop and manufacture the XLR10-RM-2 (the official designation of the Viking rocket engine) was signed in 1946, and the first of ten engines was successfully test fired a year later. Although the early Viking launches were not impressive, in time the rocket exceeded the altitude capabilities of the captured German V2s.

The engine consisted of a large combustion chamber, a turbopump unit, and associated valves and controls. All tankage, wiring and piping were supplied by the Glenn L. Martin Company. Over the years, improvements were made to the engine. Thus, it developed 20,450 lb (9,280 kg) thrust on the first Viking flights, 21,429 lb (9,720 kg) on the last. The thrust could be varied over rather wide limits by merely altering the feed pressure. It was gimbal mounted for directional control.

The thrust cylinder consisted of the combustion chamber and a venturi nozzle. It was composed of a sheet nickel inner jacket rolled into a cylinder and welded. The motor was the first to use pure nickel for both liner and nozzle. According to Robertson Youngquist, nickel was chosen because of its superior conductivity (thermal) and higher melting point, the 20,000 lb (9,100 kg) motor was large enough to permit fabrication from sheet stock, the only form of nickel available at the time.

The engine was cooled by circulation of the alcohol in an outer jacket surrounding the combustion chamber, supported approximately a 0.25 in. (0.63 cm) distance from the outside diameter of the chamber by spacer rods (that extended from one end of the chamber to the other in several short lengths). The outer jacket contour was interrupted by expansion channels which, when pressure was applied, tended to expand lengthwise to allow for thermal expansion of the combustion chamber in operation. A reinforcing ring was welded to the jacket throat section to prevent the possible distortion of the nozzle throat. The fuel entered the engine by four 1.25 in. (3.2 cm) fittings at the nozzle end of the jacket and was distributed uniformly around the cylinder.

Figure 6 Launch of Viking No. 4 rocket from the deck of the U.S.S. *Norton Sound*, 11 May 1950. U.S. Navy photo from the Andrew G. Haley Collection, National Air and Space Museum.

The power transmission and pumping units consisted of a single stage turbine and two centrifugal pumps. Hydrogen peroxide entered a gas generator containing a manganese dioxide catalyst. The resultant chemical reaction produced super-heated steam that was supplied to the turbine. The turbine operated at approximately 10,000 rpm and developed an output of approximately 300 hp. Impellers at both ends of the turbopump forced the propellants through the outlet ports of the pumps at the proper discharge pressures.

Among the key personnel involved in the XLR10-RM-2 engine program, Edward A. Neu, Jr., was responsible for thrust chamber and injector design, Albert Thatcher for the hydrogen peroxide-driven turbopump, and M. E. Parker for valve and control work (which was borrowed in large part from the MX-774 development).

Under pressure from both the Navy and the Glenn L. Martin Co., RMI Chief Engineer William P. Munger assigned Parker—in the latter's words—"to pull it all together and get off the cold flow and hot test series." These completed, on 21 September 1948, following a prototype engine firing that produced 21,000 lb (9,525 kg) of thrust for 66 seconds, the Naval Research Laboratory announced its approval. At about this time, Parker resumed his experimental project activities, and the XLR10-RM-2 production program was continued by Delacy Ferris and Henry Jatczak. The first production engine was then shipped from RMI to Martin soon after. Meanwhile, the Viking No. 1 airframe was being fabricated there during the autumn months, and by December it had been completed. The next month, the rocket was sent from Martin's Baltimore plant to White Sands Proving Ground, where preparations for static and flight firings got underway.

At the request of Milton Rosen, who directed Viking's development at the Naval Research Laboratory, Parker was sent out to White Sands Proving Ground for the first launch, which was to be preceded by a 30-second firing on the launch pad. After a dry run on 28 February 1949, the first static test took place on 11 March, following several postponements. Steam from the turbine exhaust started a fire in the tail section's pipe insulation, but the rocket was not damaged.

After the fire that occurred during the static test, Parker, John Cray, and the RMI crew worked night and day along with Youngquist and the Martin crew, in an attempt to eliminate the steam leak by tightening the bolts on the turbine and housing. They claimed success and that the gasket was intact. Rosen insisted on another static test to demonstrate engine integrity before he would allow the launch to proceed. But Parker and J. Preston Layton from Martin, confident that it was not necessary, refused the test, preferring that the rocket came down for disassembly and inspection. Weeks later, inspection at disassembly revealed that the on-pad repairs by RMI had successfully accomplished what Parker and Layton had anticipated.

The second static test, after reassembly and erection on the pad, took place on 28 April 1949. Testing was terminated prematurely when someone thought he saw smoke. However, inspection revealed no evidence of fire.

Preparations continued, and on 3 May 1949 at 0900 hours, the Viking made its maiden flight. Launch was successful and the rocket reached an altitude of slightly more than 50 miles (80 km), 163 seconds after launch. All agreed that it was a highly creditable performance for the first flight of any rocket. It fell to Earth about 10 miles (16 km) from its launch point, having stayed aloft some six minutes. American high-altitude sounding had come of age.

Since the experience of early flights was applied to the design of later models, no two Vikings were exactly alike. Flights 2 and 3, which took place in September 1949 and February 1950, reached approximately the same altitude as the first. *Viking 4* was fired on 11 May 1950 from the U.S.S. *Norton Sound* in the Pacific Ocean. The test gave new information about the problems of launching rockets at sea and allowed scientists to gather data on cosmic rays in the upper atmosphere at the geomagnetic equator. Despite the difficulties of a shipboard launch, the Project Reach (as the operation was called) rocket soared to a height of 106 miles (170 km), very nearly the theoretic maximum of that particular vehicle.

The next three flights were made from White Sands Proving Ground. *Viking 5* reached an altitude of 108 miles (174 km) in November 1950. The following month, *Viking 6* reached only 4 miles (6 km) because of a rapid deterioration of aerodynamic stability after fin control was lost. *Viking 7*, flown in August 1951, established a record height of 136 miles (219 km). All three carried instruments that included Geiger counters, ionization chambers, photomultiplier tubes and cameras.

More than a year went by before the next firing. The time was used to carry on design work, in process since 1950, for a larger, heavier Viking with a motor capable of firing for more than 100 seconds (compared to the 50 to 80 seconds of the earlier models). Almost all the extra 3,500 lb (1,600 kg) weight going into the new Vikings was propellant, not structure. The modified rocket could go higher than the old Vikings or carry heavier payloads to the same altitude.

Seven of the new Vikings and their rocket engines were built and flown. The only failure was the first one, *Viking 8*, which experienced one of the most unusual accidents in rocket history. During static testing on the morning of 6 June 1952, it broke loose from the test stand, flew 4 miles (6 km) up, and crashed on the desert 5 miles (8 km) away. Scientists and engineers on the scene could hardly believe their eyes as the rocket rose into the air. On future models, four tie-down points, rather than two, were used, and the rocket tail section was strengthened.

The remaining six Vikings were fired between 15 December 1952 and 1 May 1957. *Viking 11* distinguished itself in May 1954 by soaring to a new record height of 158 miles (254 km) with 852 lb (286 kg) of instruments aboard. Its RMI rocket engine fired for 103 seconds and brought the vehicle to a velocity of 4,300 miles per hour (1,691 km/h), also a record for the series.

Viking 12 was, in many ways, the last true member of the family. It reached a height of 144 miles (232 km) on 4 February 1955. The last two Vikings were fired, not for high-altitude research, but to check out components for the forthcoming Vanguard space carrier vehicle. They were flown from Cape Canaveral, Florida, rather than from White Sands Proving Ground. Homer E. Newell, Jr., head of the Naval Research Laboratory's Rocket Sound Research Section, reflected on Viking from the perspective of several decades:

> . . . the Viking, although of a marvelous design . . . found very little use. The dozen rockets bought for the development program were, of course, instrumented for high-altitude research. But Viking was too expensive. The groups engaged in rocket sounding each had perhaps a few hundred thousand dollars a year to expend on the research and a single Viking would have eaten up the whole budget. . . . It had been hoped that Viking would be much less expensive, but before the end of the development these rockets became almost as expensive as new V2s. So Viking found no takers among the atmospheric sounding groups and would probably have been shelved had it not been chosen as the starting point for the Vanguard IGY [International Geophysical Year] satellite launching vehicle.

His assessment may not be entirely fair. True, Viking was an expensive vehicle, but it allowed valuable research on the nature of upper atmospheric phenomena. Even more important, however much it may have owed to imported V-2 technology, Viking was an American project that provided vital experience to a growing cadre of American rocket scientists and technicians. This experience would prove its worth in the years to come—for RMI, Glenn L. Martin Company and the Navy.[29]

Other Important Technical Developments

The RMI story would hardly be complete without briefly relating some pioneering development work that made major inroads into the state-of-the-art that have lasted to the present. These developments are represented by the following engines:

1. 1949-1951 XLR22-RM-2:
   Bleed Turbine Turbo Rocket, 5,000 lb (2,270 kg) thrust, using liquid ammonia LOX.
2. 1950-1951 XLR26-RM-2:
   Topping Turbine Turbo Rocket, 5,000 lb (2,270 kg) thrust, using white fuming nitric acid and kerosene.
3. 1951-1953 XLR30-RM-2:
   Super Viking Rocket, 50,000 lb (22,700 kg) thrust, using liquid ammonia and LOX.

These engines were Navy-sponsored experimental developments that never reached application stage, but which demonstrated highly integrated engines of extremely low weight. They operated at high chamber pressures, high turbine speeds and utilized advanced engine cycles. The topping turbine, in fact, had the regeneratively-cooled turbine blades located just forward of the throat in the main thrust chamber.

The bleed chamber turbine drew gases from the main chamber and diluted them with fuel to bring the temperature down to 1200°C for the turbine drive. The bleed turbine experimental work led to the development of components for the XLR30 Super Viking engine, later to the LR99 for the X-15.

A most important development during this period was the "Spaghetti" construction, as it was dubbed by its inventor, Edward A. Neu, Jr. This concept, upon which Neu applied for a patent in 1947, used tubes preformed in such a way that, when properly bundled, they formed the shape of the combustion chamber. The longitudinally-placed tubes welded together thus served the dual function of being the chamber itself and the coolant passages. This chamber was also very lightweight yet strong.

[29] All material in this section was drawn from Milton W. Rosen, *The Viking Rocket Story* (New York: Harper and Brothers, 1953), passim; Milton W. Rosen, "The Viking Rocket; Memoir," *AAS History Series*, Vol. 7, Part II (San Diego: American Astronautical Society, 1986), pp. 429-443, paper presented at the 6th International Astronautical Federation History Symposium, Vienna, October 1972; Homer E. Newell, Jr., *High Altitude Rocket Research* (New York: Academic Press, 1954), passim; Homer E. Newell Jr., *Beyond the Atmosphere: The Early Years of Space Science* (Washington, D.C.: National Aeronautics and Space Administration, 1980), p. 58; and interviews with Robertson Youngquist by Ordway, 27 August and 29 December 1982.

The concept, now taken for granted, was the first tried on a nozzle section of an RMI 400 lb (180 kg) thrust chamber. The tubes were of copper and joined by silver solder. Lou Arata made the first one himself, including the die that formed the tubes into the proper profile. Bud Parker, then running the company's experimental project group, was impressed by the idea and, with help from Neu and Arata, applied it to both the XLR22 (with Bob Cramer leading the effort) and the XLR26 (with Neil O'Rourke as the project engineer).

During this early phase of the "Spaghetti" technique development, the tubing used ran a gamut of possibilities—including copper, nickel, stainless steel (347), inconel, aluminum and carbon steel. Thrust chambers of each of these metals were fired successfully. Methods of joining the tubes varied from soft solder, silver solder, welding and furnace brazing. All of these methods were also found to be successful, with furnace brazing considered the optimum. It was later incorporated into the X-15's LR99 engine.

The "Spaghetti" technique found its way into all the engines mentioned above, the XLR22-RM-2, XLR26-RM-2 and XLR30-RM-2. This method of construction soon spread throughout the industry by the mid-1950s, for example: Rocketdyne with its Jupiter and Atlas engines and Aerojet with its Titan engines. Eventually, all of Rocketdyne's powerplants for the Saturn series used some variation of Ed Neu's tabular construction. Today, the Space Shuttle's main engines are built upon this concept and also utilize an engine cycle related to the topping system that RMI pioneered during 1950-1951.

Credit must be given here, too, to RMI's chemistry laboratory, founded in 1946 and guided until 1953 by the Austrian-born chemist Dr. Paul F. Winternitz. Winternitz conducted an extensive propellant research program and was responsible for perfecting the liquid ammonia-LOX combination which, as noted above, powered the XLR22-RM-2, the XLR30-RM-2 and, later, the X-15's LR99. Winternitz probably made the first performance calculations on the boranes as rocket fuel from 1947, and he also studied ethylene oxide as a monopropellant and other exotic high-energy mixtures. (Interestingly, Winternitz's experience with LOX went back to pre-World War I, when he was a research chemist with the Linde Corporation in Bavaria; during 1915-1919, as a lieutenant in the Austrian Army, he worked on explosives, especially those using LOX.)[30]

Miscellaneous Projects

Throughout its existence RMI proved its innovativeness. Apart from projects such as JATOs, the famous 6000C4 and the Viking powerplants, RMI also undertook several smaller but no less important endeavors.

The RMI logbook cited above reports a moment of inertia test on 1 May 1944 with a 60 lb (27 kg) autogiro blade. According to Lou Arata, this started out as Pierce's project. Shesta recalled the blade came from Pitcairn Aviation, Inc., and was an attempt to test the feasibility of a liquid-propellant, rocket-powered helicopter (Figure 7). Chuck Towns managed the program.

[30] All information in this section from typed M. E. (Bud) Parker notes, n.d. in "Reaction Motors, Inc." file, National Air and Space Museum.

For the purpose, said Arata, an A-frame was secured to a concrete slab. The rotating mechanism was affixed in the middle with a 2 in. (5 cm) diameter tube protruding from it. This tube simulated the blade for the combustion tests. Two 50 lb (22 kg) LOX/gasoline, regeneratively cooled rockets were attached to each end of the tube. The fuel and oxidizer feed lines led down the tube to the center and led from their respective tanks.

Figure 7 Rotorcraft RN-1 "Pinwheel" helicopter with Reaction Motors, Inc. 16F1 hydrogen peroxide 16-lb thrust rocket engines on the rotor tips, Glendale, California, ca. 1951. Courtesy of Frederick I. Ordway III.

The blade worked—after a fashion. It rotated as it should have done, but Arata and Shesta said there was a great leakage of LOX. The LOX also evaporated and the feed was intermittent. The project was abandoned soon afterwards, though it may have been the first attempt to build a liquid-propellant rocket helicopter. The experiment anticipated RMI's Rocked-On-Rotor (ROR) project of 1954, which was much more successful.[31]

[31] RMI Logbook; and Arata interview.

In July 1944, Shesta recalled, "our engineers were summoned to the [Navy's] Bureau of Aeronautics in Washington for consultation in connection with the German V-1." "At the same time," he continued:

> certain sketchy details of the construction of the powerplant were relayed to us and we were requested to construct and test a duplicate thereof. Work on this project was started in the middle of July and was completed on or about the first week in August 1944. These tests continued to about the middle of September 1944, at which time we took the motor to the Philadelphia Naval Aircraft Factory in Philadelphia for further testing.

The RMI logbook reports one static test at the New Jersey plant on 15 September 1944:

> Made 5 test starts without German reg.[ulator] . . . Ran as long as air was injected. Put in reg. made 3 test starts. #3 kept unit firing without air injection. After about 30 sec.[onds] the impulsing smoothed out and increased in intensity.

The V-1 engine was an air-breathing pulsejet, not a rocket; but it did fit the mission of the company: the development of reaction motors. The 1946 GALCIT survey, Jet Propulsion, noted RMI's little-known contributions in this unique area of reaction propulsion. RMI "successfully modified the V-1's aeroresonator to extend the valve life to approximately four times that [of] the German valve," the survey reported. The company "has practically completed development of an air-fuel intake system which will greatly reduce the draft of the propulsion device and increase the efficiency," it added.

Arata made a V-1 pulsejet during the course of this research; and initial tests were also made with scaled-down versions subjected to wind-tunnel tests that used an antiquated aircraft engine for ramming in the air. In other tests, simulated air was fed from tanks containing 21% oxygen and 79% nitrogen. John Shesta briefly summed up the purpose for this interesting departure from RMI's rocket work. The Navy, he said, was exploring the possibilities of producing its own pilotless V-1 missile. The pulsejet-powered Navy Loon missile did emerge following World War II, but the motor was manufactured by the Ford Motor Company.[32]

Another unusual RMI project was a rocket-driven boat. G. Edward Pendray, in his book *The Coming Age of Rocket Power* (1945), states that this vessel was tested during February and March 1944 for several purposes, including the delivery of mail from ship-to-shore, rescue work, and "for sport events." Wartime restrictions apparently did not permit the RMI founders and former fellow members of the American Rocket Society to reveal to Pendray the primary purpose of the boat: to act as a very fast beach landing craft in the projected invasion of Japan (Figure 8).

Pendray did say that the boat was powered by a standard RMI 250 lb (113 kg) thrust motor using LOX and gasoline. Lou Arata believed that it was one of the Lark missile engines, although the Larks employed nitric acid and aniline. Since John Shesta

---

[32] *Ibid.*; John Shesta, untitled document unpublished in the RMI archives, Thiokol Corp., Newton, Pennsylvania (now Morton-Thiokol, Chicago, Illinois).

remembered that the boat's rocket engine "burned a very long time, about five minutes," it could well have been a Lark sustainer that ran for 260 seconds. Pendray may not have been informed of the hypergolic acid-aniline fuel and was only personally acquainted with LOX and gasoline, which powered all the ARS engines. At any rate, Franklin Pierce made several experimental runs of the boat on the Severn River, near Annapolis, clearly indicating Navy interest in the project. Speeds up to 40 mph (64 km/hr) were attained with "no problem;" but steering was found to be extremely difficult especially at higher speeds. Nothing further was heard of the curious rocket boat, which never reached full development stage.[33]

Figure 8 Experimental rocket-propelled landing craft of Reaction Motors, Inc, using a standard 250-lb thrust liquid JATO, preparing for test on the Severn River, near Annapolis, Maryland, during February-March, 1944. Photo from John Shesta. H. F. Pierce appears to be on the right (cigarette in mouth). Courtesy of Frederick I. Ordway III.

On a spare-time, "for fun only" basis, RMI Chief Test Supervisor, Ernest John "Buck" Pellington, a former World War II B-24 crew chief with 58 missions over Germany, supervised the design and construction of a stainless steel tubing, rocket-propelled ice sled. Fitted with a 400 lb (180 kg) thrust LOX and alcohol engine, the 1,648 lb (747

---

[33] G. Edward Pendray, *The Coming Age of Rocket Power* (New York: Harper and Brothers, 1945), pp. 171-72.

kg), 16 ft (4.8 m) long sled attained a top speed of 90 mph (152 km/hr) in its initial test run on frozen Lake Hopatcong near the RMI plant at Dover, New Jersey. Pellington encountered rough ice and steering difficulties and had to shut down the engine after 18 seconds. Otherwise, greater velocities for the streamlined, aircraft-like sled would have been possible. The Austrian Max Valier and the American Harry Bull made and rode rocket sleds in 1929 and 1931, respectively, but this was the first liquid-propellant type.[34]

RMI became involved with another rocket-powered helicopter in 1951. The firm made the 16F1 hydrogen peroxide engines (16 lb, 7 kg thrust) for the tiny 100 lb (45 kg), two-bladed, strap-on helicopter invented by Gilbert Magill. Magill manufactured the machine on an experimental basis under an Office of Naval Research contract through his Rotor-Craft Corporation of Glendale, California. The rockets were attached to each blade. The helicopter, known as the Rotorcraft RH-1 Project Pinwheel, was strapped to a man's back and waist with the man resting on a bicycle seat. The military implications of Magill's invention were many and included soldiers leap-frogging over entire armies, climbing inaccessible terrain, etc. In effect, this was one of the first so-called "jet packs," though Magill's project seems to have been short-lived.[35]

A more bonafide rocket helicopter was RMI's (Rocket-on-Rotor) produced in 1954, though preliminary work was begun in 1951, perhaps as a follow-on to the Magill Pinwheel. It consisted of small XLR32-RM-2 40 lb (18 kg) thrust hydrogen peroxide motors mounted on the three blade tips of a full-sized helicopter, the Sikorsky HRS-2 (S-55 type) of the U.S. Marine Corps. The system weighed 67 lb (30 kg) dry, yet could increase a helicopter's lifting potential by 20%; which was a great advantage in wartime situations. A dome-shaped tank above the rotor hub carried enough fuel for six minutes, or 20-25 take-offs. Tests covered 1,000 starts and stops with helicopters of the Marine Experimental Squadron HMX-1 at Quantico, Virginia. A 99.5% reliability record was achieved. Faster take-offs, high altitude, and improved rates of climb and hovering ceiling were also advantages of the ROR; yet the system did not become operational.[36]

Apart from these miscellaneous projects, RMI was involved with many other, often fruitful, lines of research during this pioneering period. For example, from 1946 the company initiated the development of gas generators for aircraft catapults using the rocket principle. This system, called ICCP (Internal Combustion Catapult Powerplant), was conducted under a Navy contract. Rocket-ramjet studies were undertaken from 1946. The X50-AP1 Auxiliary Power Pack, designed under an Air Force contract to produce auxiliary turbine power operating independently of the atmosphere, was devel-

[34] I. Essers, *Max Valier, A Pioneer of Space Travel* (Washington, D.C.: National Aeronautics and Space Administration, 1976); "Rocket Propelled Boat Tested at Lake Hopatcong," *Morristown Daily Record* (Morristown, New Jersey), 3 March 1947; and "Midvale Vet Hopes to Get Jet-Iceboat Up to 200 MPH," Midvale, New Jersey, paper unknown, March 1947, from "Reaction Motors, Inc." file, National Air and Space Museum.

[35] "Tiny Rocket Helicopter Tested Here," *Los Angeles News*, 23 October 1951.

[36] "Marines Visit RMI Employees, Show Off ROR," *RMI Rocket*, 7 March 1956, pp. 1,8; "Tiny New Rocket Engine Boost Payload for Copters," *Aviation Week*, 27 September 1954, p. 17; and "Rocket Assistance for Helicopters," *Flight*, 66, 24 September 1954, p. 500.

oped from 1950. A solid propellant gas generator for tank pressurization was manufactured in 1951. In the same year, RMI evaluated ceramics for thrust chambers, gas generators and igniters. A more exotic project was the rocket for drilling into the ground, a rocket-age version of the pneumatic drill that appeared in 1951. Boundary layer control air-pumping systems studies were made for the Chase and Cessna Aircraft companies in 1952.[37]

In 1954, the US Air Force conducted a competition for a second source of the Atlas missile's three-engine principal propulsion system under development by Rocketdyne. The competition was important, for it was widely believed at the time that there was room for only two contractors in the business of developing and manufacturing very large liquid-propellant rocket engines. By virtue of its contract for the Atlas missile, Rocketdyne was one of these.

Reaction Motors and Aerojet went all out to become the second beneficiary of the eagerly sought contract. RMI designed and built a full-scale motor with a nominal thrust of 185,000 lb (83,900 kg), using spaghetti construction along with a prototype full-scale, two-stage turbopump, to give the company a head start in the event it were to win the competition. Meanwhile, Aerojet focused its efforts and money principally on new test facilities in Sacramento, California. As it turned out, Aerojet won the competition with the understanding that they would subcontract an important portion of the Air Force work to Reaction Motors—RMI's proposal, though weak in the facilities area, was strong technically.

Thus, RMI found itself devoting an increasing effort in supporting Aerojet ICBM work. As time progressed, RMI specialized more and more in Vernier engine technology, important to the Atlas ICBM configuration.

"Small engines they may have been," observed Bud Parker, "but at least they represented a discreet package." But problems arose; though Aerojet clearly wanted to take advantage of Reaction Motors' know-how and technology, it also began to explore ways of eliminating RMI as a competitor. In this, it was at least partially successful.

In the mid-1950s, the Air Force established a requirement for a backup ICBM to the Atlas. Aerojet's responsibilities soon encompassed the primary propulsion for this missile—and Titan, unlike Atlas, did not incorporate Vernier engines. Recognizing the spot RMI was clearly in, management immediately began to seek new work divorced from the ICBM program. By the time Titan became an approved project in January 1957, RMI was out of the ICBM game.

Reaction Motors had moved swiftly, and during 1956 a number of other programs came into being. One was known as ARA for Advanced Research Airplane, the beginning of the development of the XLR-99 for the X-15. Another was the development of the XLR-44, an integrated packaged liquid propellant engine for the Sparrow air-to-air missile. It grew out of the RMI "Can Do" project, a company-sponsored effort during 1953 and 1954 to demonstrate the feasibility of this packaged liquid engine using solid-propellant tank pressurization. The XLR-44 was brought through the development cycle and became fully qualified during the 1957-1958 period. Production was limited, how-

---

[37] Interview with M. W. (Bud) Parker by Ordway, Huntsville, Alabama, 26 March 1982 and subsequent conversations, and letter from same, 27 July 1982; Shesta interviews and Youngquist interviews.

ever, and the packaged liquid version of the Sparrow missile was never made operational. However, during the same period the LR-58 Bullpup engine was successfully developed and qualified, and went into large-scale production. This project will be discussed in the third and fourth chapters that appear in Volume 7 of this subseries edited by John L. Sloop (Volume 12, *AAS History Series*), covering the Reaction Motors Division of Thiokol.

Still another area probed by Reaction Motors during the final years of its independent existence was the Super P, a development program calling for a hydrogen peroxide oxidizer/JP4 fuel turborocket. Designated XCR-40, it was to be employed in a high-performance Navy fighter aircraft. Super P was a highly integrated, light-weight unit using an aluminum spaghetti combustion chamber. The engine was fully developed and had approached the final stage of delivery to the customer, Chance Vought, when an unfortunate accident in an RMI test stand killed one of their employees. Safety concerns led to major redesigns of the turbopump. In the end, delays brought on by the redesign effort effectively killed the program.

Other activities included a 500,000 lb (nearly 227,000 kg) thrust JP4 fuel, hydrogen peroxide oxidizer pressured rocket engine for a U.S. Air Force supersonic sled. It was, however, canceled after engine design was completed. Through 1957, the company was engaged on the Corvus, a U.S. Navy-sponsored, Temco-built missile program, that resulted in a development contract for the XCR-48 turbo-rocket engine; it operated on IRFNA and MAF (mixed amine fuel) propellant (hydrazine). The program proceeded through engine qualification. Another effort was the development of a 35,000 lb (15,870 kg) thrust test stand unit for the Army Ballistic Missile Agency's nosecone test activities. The unit was designated 164-MHG Burner.

Finally, during the 1953-1957 period, Reaction Motors carried on some solid propellant research and development. According to Parker, "a very significant solid propellant research group was established, and this group became well-known in their efforts to leap frog into the advanced, high-performance solid propellant area." RMI's solid activities were closely coordinated with the Olin Mathieson Chemical Corporation, and they were rather successful in pursuing contracts with the Air Force. Some work went on in the hybrid propellant area as well.[38]

---

[38] Interview with M. W. (Bud) Parker by Ordway, Huntsville, Alabama, 31 August 1984 and letter from same with notes, 20 September 1984.

Chapter 9

Early Experiments with Erosive Burning in Solid Rockets[1]

J. M. Murphy[2] and J. R. Osborn[3]

In the years immediately following the Second World War, considerable research and development attention was focused upon the liquid rocket motor. At that time, except for JPL, little effort was expended in developing the solid rocket, primarily due to its comparatively low performance and its reputation for hazardous operation. However, in the early and mid-1950s, with the advent of JPL-developed composites having the newer binder systems, the solid rocket motor was recognized as a viable alternative to, and replacement for, the liquid rocket motor in several applications. As a result, the research and development efforts in the area of solid rocket propulsion expanded, resulting in an increase in performance and in the number of problems unique to the solid rocket motor.

One of these problems, that of erosive burning, was particularly perplexing since at that time there was really very little understanding of the normal combustion processes for the castable composite propellant. Recognizing a need for rudimentary information regarding the mechanisms affecting erosive burning, a research group at Purdue University in 1960, under the direction of M. J. Zucrow, Professor of Gas Turbines and Jet Propulsion, developed a new and unique technique for instantaneously measuring the burning rate of a propellant under erosive conditions. Funding for the program was obtained with a grant from the Air Force's Rocket Propulsion Laboratory.

---

[1] Presented at the Fifteenth History Symposium of the International Academy of Astronautics, Rome, Italy, 1981.

[2] Space and Electronics Systems Division, Martin Marietta Aerospace, Denver, Colorado.

[3] School of Aeronautics and Astronautics, Purdue University, West Lafayette, Indiana.

Previously, erosive burning rates had been determined using the interrupted burning technique. It involved terminating the thrust of a solid rocket motor having an initial port to throat area ratio near unity, and determining an average burning rate from a measurement of the initial and final webs. This technique could only determine average values of erosive burning rates over a range of relatively low subsonic Mach numbers.

The Purdue technique overcame this difficulty by constructing a gas generator and separate test section, in which the erosive burning measurements were made from observations of the burning surface in the test section. The gas generator was a solid rocket motor having a high mass flow, while the test section was two dimensional having only a small amount of propellant. A later development in the program involved moving the propellant surfaces in the test section inward at a velocity equal to the burning rate by means of a closed loop servomechanism. As a result, the burning rate was measured instantaneously, while the Mach number in the test section remained constant, since the port area was maintained constant by the inward motion of the burning surfaces. Also, it was possible to obtain erosive burning data near Mach numbers of unity, since the port to throat ratio could be maintained equal to unity. The experimental technique described above has been used extensively by almost all other experimenters since its development at the Purdue University Rocket Laboratory.

The experimental program involved determining the erosive burning constants for composite propellants, as well as for some double base propellants. Mainstream velocities in the test section ranged from 600 to 3,600 feet per second, and the combustion pressure ranged from 400 to 600 psi. The threshold phenomenon was observed for the composite propellants, and its value was determined for the range of pressures described above.

The paper also described the early experimental facilities including the servomechanism technique. It also presented the pioneering work of the solid rocket propellant group at the Purdue University Rocket Laboratory, which outlined the way for the high performance, nozzleless solid rocket motors of today. Accurate knowledge of the regression rate of a solid propellant is essential to the design of high performance solid propellant rockets, since both the thrust level and the burning time depend on the regression rate of the propellant. In the early days of research into the combustion of solid propellant rockets it was clear that no experimental data existed for predicting the regression rates of solid propellants. Thus, the performance of newly designed solid rocket motors had to be substantiated by numerous firings of full-scale rocket motors.

The combustion of a solid propellant within a rocket motor is complicated, when products of combustion continually sweep the receding surface and alter the normal ballistic behavior of the propellant. This deviation from normal ballistic behavior, which is known as erosive burning, is important at high subsonic combustion gas Mach numbers, since the regression rate under these conditions may be two to three times the normal linear burning rate of the propellant.

In general, it is desirable, because of weight considerations, that the volumetric loading fraction of a solid rocket motor approach unity. In order to achieve a high volumetric loading fraction, the ratio of the initial port area of the solid propellant grain to the throat area of the nozzle should approach unity. This small port to throat area ratio provides for high subsonic gas velocities in the port of the solid propellant motor.

The most severe difficulty in studying the phenomenon of erosive combustion was the absence of a sufficient volume of reliable and systematic experimental data for propellants burning under conditions which simulate the conditions that are found in solid rocket motor.

As a result, the Purdue group devised two new and unique experimental measurement techniques to provide methods to obtain accurately the regression rate of a solid propellant which is burning under conditions that simulate actual rocket motor conditions. The first method, devised about 1960, involved photographing the receding propellant surface in a special test section downstream of a gas generator. The second method, developed a few years later, involved the use of a servomechanism for maintaining the receding propellant surface at a fixed position. In almost every instance later, researchers have utilized the concepts developed by the Purdue group in designing their experimental erosive burning apparatus.

The Photographic Technique

The experiments were conducted with two-dimensional rocket motors, comprising a gas generator and a test section; the propellant sample was burned in a two-dimensional test section. Figure 1 illustrates the essential features of the test section; the propellant samples were bonded to its diverging sides. The divergence of the latter was such that the axial velocity of the combustion gases (discharged from the gas generator) was substantially constant in the region where they wetted the propellant samples.

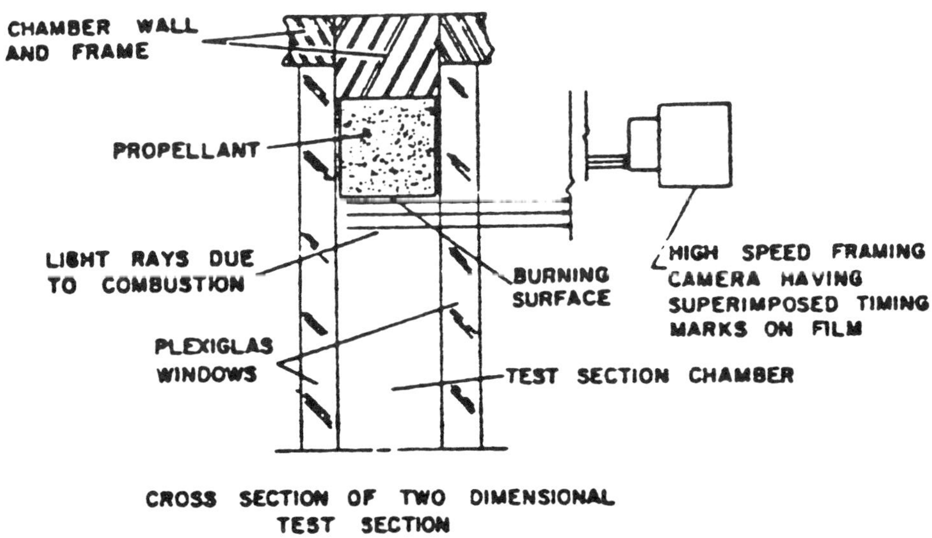

Figure 1 Schematic Diagram of Photographic Method for Determining the Burning Rate of a Solid Propellant Sample.

In each experiment, the propellant sample burned in the test section was taken from the same batch of propellant that was burned in the gas generator. Consequently, there was no significant difference in the composition of the gases produced by the gas generator and those wetting the surface of the burning propellant sample located in the test section.

The test section was equipped with transparent Plexiglas (Rohm and Haas) windows, so that light could be transmitted through the combustion chamber. A sample of the propellant to be investigated was mounted on the upper surface, and another sample was mounted on the lower surface of the test section; each sample was in the form of a parallelepiped block, 5 in. long, 1 in. wide, and 1/2 in. thick. The sides of the upper propellant sample were visible through the aforementioned windows.

Only the upper sample was photographed during the burning period; a 16-mm Fastax (model WF3) high-speed framing camera was employed for that purpose. The camera, which was equipped with a Wollensak 152-mm f/4.5 cine telephoto lens, had a field of view approximately 1.5 in. in diameter. The camera was located with its optical axis normal to the plane of the transparent window. The average speed of the camera was approximately 720 frames/sec, and the speed was adjusted for each static firing of the research rocket motor; the aperture setting was f/32. All of the photographs were taken with Ansco Type 229 (color) film. A time base on the film was provided by a small argon lamp flashing at 120 cycles per second (cps).

The combustion chamber pressure measurements were made by conventional electrical pressure transducers (Wiancko P 1951) and an oscillograph (Consolidated Electrodynamics Corporation Type 5-114-P4). The photographic record, and the pressure traces recorded by the oscillograph, were synchronized by employing separate event markers for recording the activation of the igniter circuit. In this way, the burning rate of the propellant sample, and its true combustion pressure, were coordinated.

A disadvantage of the photographic technique was that the burning rate had to be inferred from the photographic data. That is, the burning rate r of a propellant sample was calculated from the recession of the burning surface determined from the high speed photographs of its combustion. The burning rate was calculated by means of the following equation. Thus,

$$r = (SF)\,\Delta x/\Delta t_b \tag{1}$$

where SF is a scale factor.

The average Mach number of the combustion gases in the test section M_{ts} was calculated from the equation:

$$\frac{A_{ts}}{A_{th}} = \frac{1}{M_{ts}}\left\{\frac{1 + (k-1)/2 \quad M_{ts}^2}{(k+1)/2}\right\}^{(k+1)/2(k-1)} \tag{2}$$

The average velocity of the combustion gases in the test section u_g was determined from the equation:

$$u_g = M_{ts}(kgRt_f)^{1/2}\left[1 + (k-1)2M_{ts}^2\right]^{-1/2} \tag{3}$$

The calculated values of the burning rate were then plotted either as a function of the combustion chamber pressure and/or the combustion gas velocity past the burning surface of the propellant.

During an experiment, the following parameters were held constant: (1) the composition of the combustion gases, (2) the combustion pressure p_c, and (3) the propellant temperature t_f. The following variables were measured: (1) the static pressure at the fore-end of the gas generator, (2) the static pressures at the entrance and the exit sections of the test section, and (3) the burning rate r of the propellant sample located in the test section.

Experimental Results Using the Photographic Technique

The experimental results were rather unique, since for the first time reliable data were obtained at very high subsonic Mach numbers. Table 1 presents the ranges over which u_g and p_c were varied during the investigation. Figure 2 presents the total burning rate r as a function of the combustion gas velocity u_g for the following constant values of combustion pressure: p_c = 400, 500, and 600 psia. The values of r measured at a given operating condition were repetitive to within 4 percent. At each operating point a minimum of two to four runs were conducted. The curves indicate that for the particular propellant there is a "threshold" velocity u_{tz} above which the observed burning rate r is faster than the linear burning rate r_o; Figure 2 shows that u_{tv} is a function of p_c.

Table 1

RANGE OF PARAMETERS

PARAMETER	**RANGE OF VARIATION**
Combustion pressure	400-600 psia
Velocity of the combustion gases	600-3000 fps
Initial temperature	60 ± 5F

Experiments were also conducted for determining the influence of the propellant temperature t_i upon the burning rate r; the latter includes the contribution due to erosive burning. Two different values of propellant temperature were investigated: $t_i = -15 \pm 5$ and $t_i = 100 \pm 5$F. Table 2 presents the results obtained from experiments for determining the influence of t_i.

The data presented in Table 2 indicated that the erosive burning rate r_e is influcnccd by the initial temperature of the propellant t_i and increases as t_i decreases. Again, for the first time reliable data were obtained for high subsonic Mach numbers for a range of pressures and initial temperatures, unusual data for the early sixties.

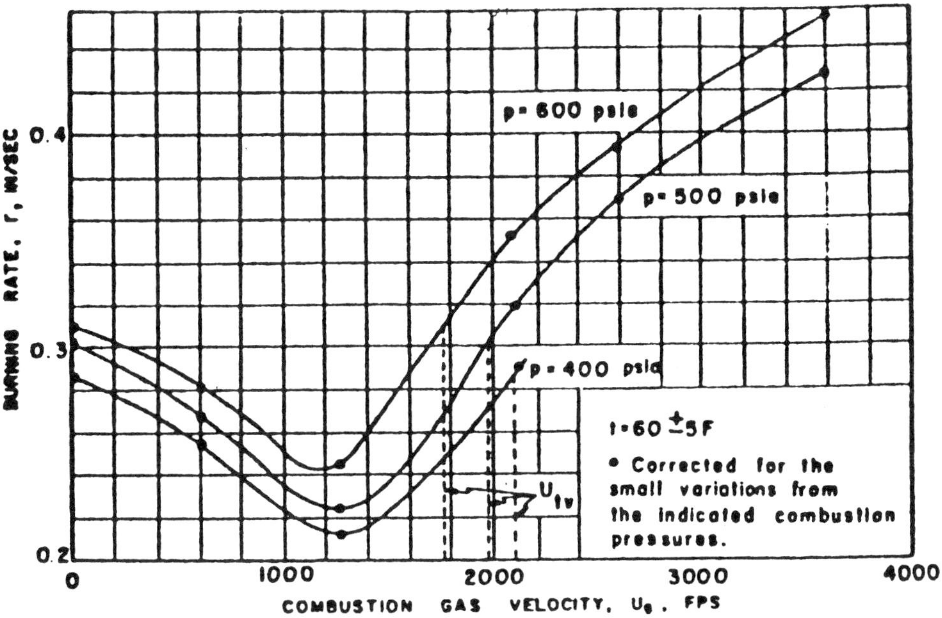

Figure 2 Effect of combustion gas velocity and combustion pressure on the total burning rate of a solid propellant.

Table 2
COMPARISON OF BURNING RATES AS A FUNCTION OF INITIAL TEMPERATURE (t_i)

PARAMETER	-15F	60F	60F	100F
Combustion pressure, psia	535	547	652	620
Velocity of the combustion gases, fps	2460	2560	3140	3180
Total burning rate r, in./sec	0.38	0.37	0.42	0.36
Linear burning rate r_o, in./sec	0.28	0.31	0.33	0.32
Erosive burning rate r_e, in./sec	0.10	0.06	0.09	0.04

The Servomechanism Technique

Recognizing the need for the development of a better method for measuring the burning rate under erosive conditions, the Purdue group developed the servomechanism technique in the early sixties. The servomechanism technique had the advantages not only of providing a direct determination of the burning rate, but also of providing a constant cross-sectional area for the high subsonic Mach numbers required in the test section. As a result, the Mach number could be maintained constant during an entire test run, and it also could be predetermined. Thus, precise, high subsonic Mach number data were obtained for the first time.

Figure 3 is a schematic of the essential features of the system described herein for measuring the burning rate of solid propellants. The propellant sample is bonded to a rectangular holder having an attached shaft, which positions the sample within a two-dimensional test section. As the burning progresses, the burning surface of the propellant sample recedes normally at a velocity r (see Figure 3). The servomechanism moves the propellant sample at a velocity r_S in a direction opposite to the receding surface. Thus, if the shaft velocity r_S equals the regression rate of the surface r the burning surface of the sample is maintained at a fixed position with respect to the test section. Therefore, the measurement of the velocity of the shaft r_S yields the instantaneous burning rate of the propellant sample.

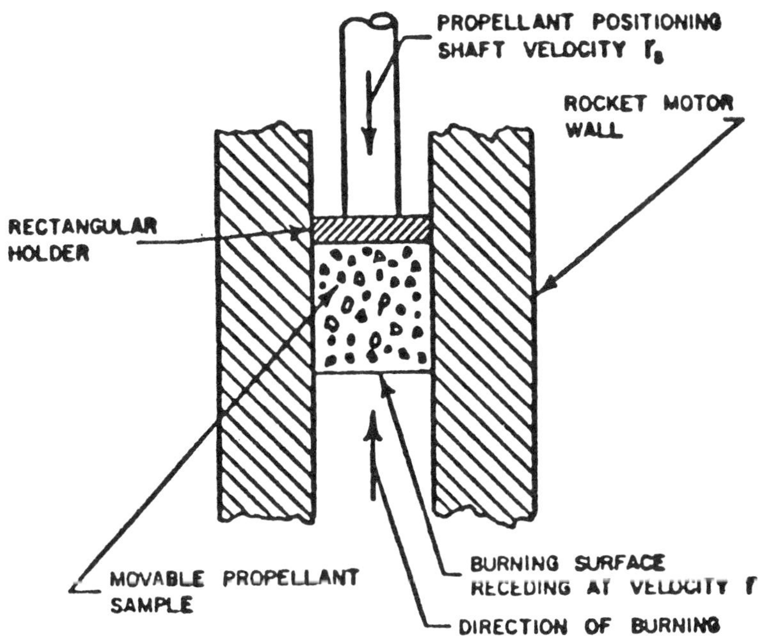

Figure 3 Burning rate measurement.

In order for the servomechanism to maintain the propellant sample at a fixed position with respect to the test section, a signal must be provided, which accurately represents the true position of the burning surface of the propellant sample. The transducer for providing that signal must be a type which utilizes some form of energy transfer rather than a mechanical measurement because of the high temperature above the propellant surface and the requirement of an undisturbed flow field for the combustion gases.

Figure 4 presents a schematic of the components of the servomechanism system. The transducer for detecting the position of the burning propellant surface is shown in the lower left-hand portion of Figure 4. The energy transfer is provided by a collimated beam of gamma rays which are emitted from a 2.5 curie source of Cobalt-60. The gamma ray beam traverses the test section and is transformed into an electrical signal, whose output provides the input to a linear ratemeter. The scintillation probe is provided with a lead shield for excluding background radiations which would produce a variable propellant position different from the true position. Incorporated into the end of the scintillation probe shield is a 0.048 in. square collimator.

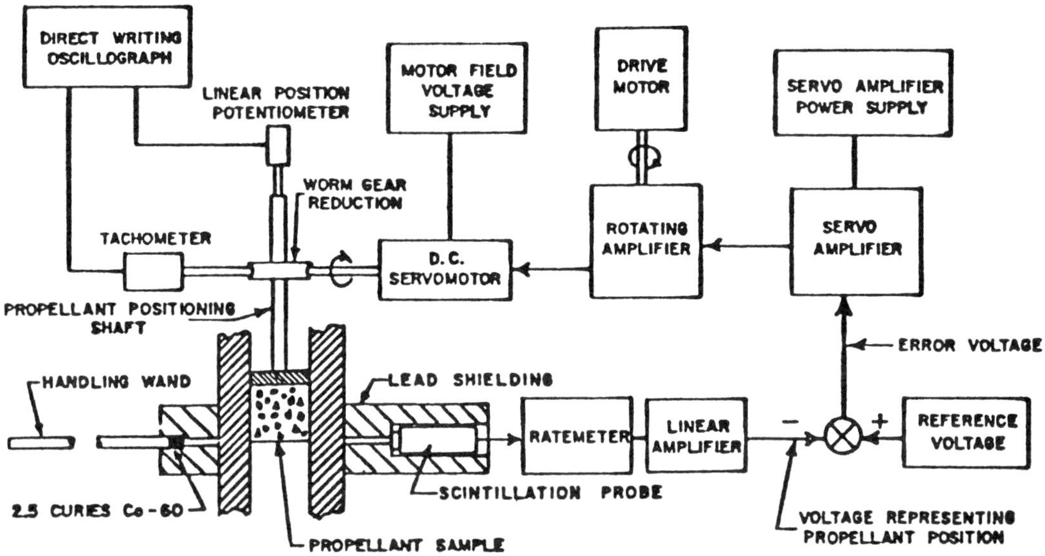

Figure 4 Servomechanism system.

The intensity of the emitted collimated beam of gamma rays that strike the scintillation probe after passing through the test section is a function of the density and thickness of the material through which the gamma rays pass. Since the walls of the test section remain in a fixed position, their attenuation is constant. In addition, the sides of the propellant sample perpendicular to the collimated beam are inhibited from burning, thereby contributing an additional constant attenuation. Thus, the emitted beam intensity is a function only of position of the propellant sample within the test section.

The collimated beam of gamma rays from the Co-60 source passes through the test section and is monitored by the scintillation probe (Baird-Atomic Model 812-B). The ratemeter (Baird-Atomic CS-400) accepts the count pulses from the scintillation probe and converts them to a counting rate in the form of a voltage which is amplified by the linear amplifier (Dana Model 2200). Thus, the output of the transducer system is a voltage proportional to the propellant surface position.

The output is then compared to a standard or reference propellant position. That reference is termed the reference voltage (provided by a constant voltage power supply, Sorenson Model QR36-4A), and it represents the desired initial position of the propellant surface. The difference between the feedback voltage and the reference voltage is called the error voltage. The error voltage essentially represents a difference between the initial or unburning propellant position and the position of the propellant surface when it was burning. A small difference in those two positions does not affect the measurement of the velocity of the surface, that is, the regression rate.

The error voltage is amplified by the servoamplifier, whose output is used to excite the field of the rotating amplifier (Diehl Model S52-A). The servoamplifier was designed and built by the Servomechanisms Laboratory of the Department of Electrical Engineering, Purdue University, to match the associated equipment.

The output of the rotating amplifier is then supplied to the armature of a 0.6 hp direct current servomotor (Diehl Model FD84-34-1). The speed of the servomotor is controlled by maintaining its field voltage constant while varying the current supplied to its armature. The variable current is provided by the rotating amplifier, and it depends upon the propellant position as was discussed above. The circular motion of the servomotor shaft is then transformed into linear motion of the propellant positioning shaft by means of a worm gear which rotates on the thread section of the propellant positioning shaft.

The regression rate or the velocity of the propellant positioning shaft is monitored by two separate transducer systems. The first system involves measuring the instantaneous position of the shaft as indicated by a linear potentiometer (Bourns Model 105) which was attached to the aforementioned shaft. The output of the potentiometer was recorded by a direct writing oscillograph (Consolidated Electrodynamic Corporation Model 5-124) which produced a shaft position-time record. Thus, the slope of the position-time trace is proportional to the regression rate.

The second system utilized the integral tachometer of the servomotor. The output of the tachometer, which was analogous to servomotor speed, was recorded by the direct writing oscillograph. Since the worm gear reduction and the feed ratio of the shaft feed mechanism were constant, the servomotor speed was directly proportional to the velocity of the propellant positioning shaft. Thus, the regression rate of the propellant sample can be read directly from the tachometer oscillogram trace.

The pressure in the test section is recorded simultaneously with the above measurements. Thus, at any instant during an experiment the regression rate is correlated with its true combustion pressure.

The Servomechanism Technique was used successfully to measure the erosive burning rates of solid rocket propellants. Figure 5 presents a schematic of the experimental apparatus which was used to simulate actual rocket motor conditions. The apparatus consisted of a gas generator which supplied combustion gases to a two-dimensional test section. Located in the test section was a propellant sample (0.970 in. wide x 5 in. long x 3 in. high) whose burning rate was determined as a function of the gas velocity and static pressure in the flow channel above the propellant sample.

Since the burning surface of the propellant sample remains fixed with respect to the flow channel, the flow area above the sample is a constant. This constant flow channel area above the sample provides for a constant combustion gas Mach number (or gas velocity) during an experiment. The feature of constant gas dynamic conditions during an experiment is the unique feature of the apparatus which is described herein.

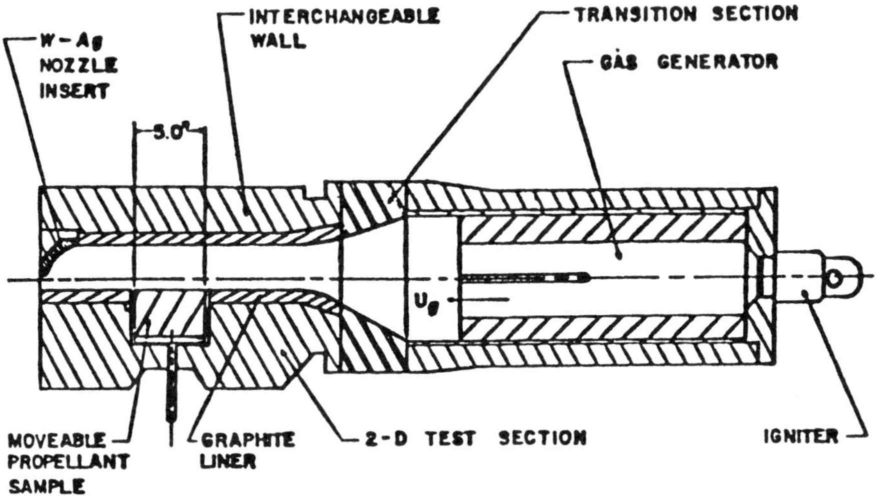

Figure 5 Erosive burning rate apparatus.

Different combustion gas velocities and static combustion pressures were obtained by the use of the interchangeable wall and nozzle insert. Four separate interchangeable walls were employed to change the flow channel area from 0.97 in.^{2} to a maximum of 3.96 in.^{2}.

To protect the test section from erosion by the combustion gases, the test section was designed to incorporate replaceable graphite liners. The graphite liners were replaced after each static firing of the apparatus. The nozzle insert was protected from the combustion gases by a tungsten 15 percent silver cap. The erosion of the graphite and nozzle cap was considered to be acceptable. The graphite liners were bonded to their respective components with RTV-88 silicone rubber.

The propellant positioning shaft and its associated gearing were designed so that the 0.6 hp servomotor would operate in its optimum speed range of 1,000 to 4,000 rpm. The worm gear reduction and threaded section of the propellant positioning shaft were designed for 0.0182 in. of shaft travel for one revolution of the servomotor shaft. Thus, regression rates in the range of 0.30 to 1.20 in. per second could be measured in the optimum speed range of the servomotor.

Experimental Results using the Servomechanism Technique

Figure 6 presents burning rate data for a typical double-base burning under erosive combustion conditions. The data are presented as the total regression rate, r_{total}, of the propellant as a function of the mass flux, G, above the propellant sample. The mass flux, G, is defined as the total mass flow rate over the propellant sample divided by the cross-sectional area of the flow channel above the sample. The data points shown in Figure 6 represent gas dynamic conditions corresponding to a gas velocity range of 790 to 2,425 feet per second and a combustion pressure range of 716 to 1,150 psia. All data were obtained for an initial propellant temperature of 70+ 5°F.

The burning rate of the propellant sample is monitored by two separate transducer systems. The first system utilizes the integral tachometer of the servomotor, which produces an electrical signal proportional to the speed of the servomotor. The motor speed is recorded on the oscillogram.

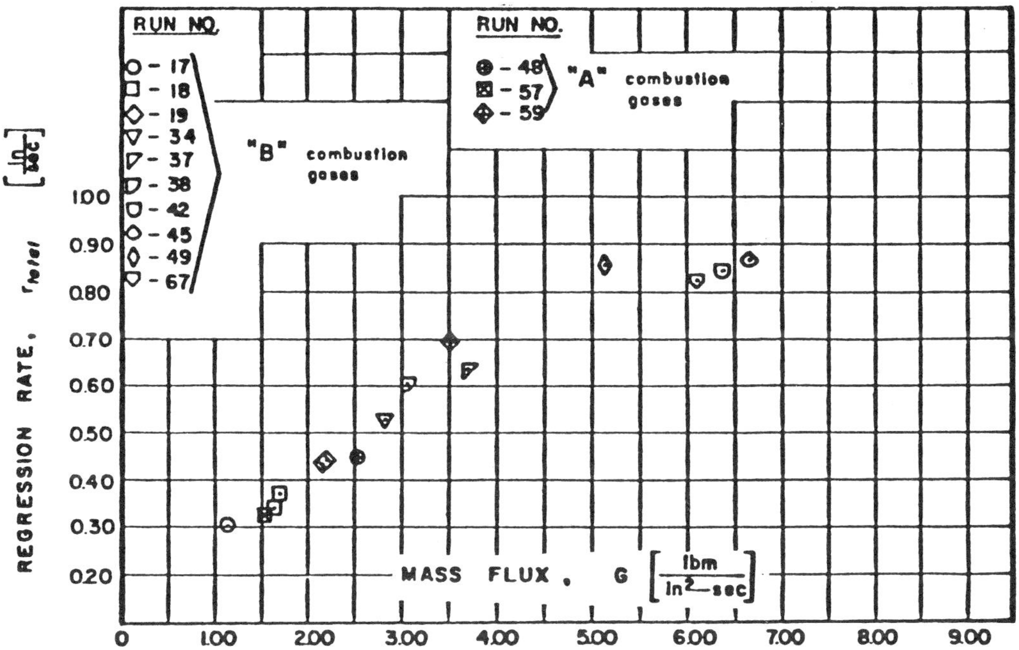

Figure 6 Erosive burning rate data.

The second means of measuring the burning rate of the propellant sample involved the use of the linear potentiometer which is attached to the propellant positioning shaft. The electrical signal which is produced by the potentiometer is recorded by the oscillograph which produces a shaft position-time trace. The burning rate measurement employing the servomotor tachometer trace was found to be the more accurate of the two methods of burning rate determination.

The value of the particular pressure parameter at the time of burning rate measurement was determined by linear interpolation of the recorded transducer output against the pre- and post-run calibrations. The static combustion pressure for the burning rate measurement was taken as the average of the three transducer readings which monitored the combustion pressure above the burning sample.

Conclusions

The pioneering work of the Purdue group in developing an apparatus for measuring the erosive burning rate phenomenon has not only provided a new and unique measurement scheme, but also outlined the way for the high performance nozzleless solid rocket motors of today.

Bibliography

Murphy, J. M., and Bethel, H. E., "Investigation of New Techniques for Measuring the Burning Rate of a Solid Rocket Propellant," Jet Propulsion Center, Purdue University, TM-62-8 (October 1962).

Osborn, J. R., and Burick, R. J., "Technique for Measuring Burning Rates of Solid Propellants," *Review of Scientific Instruments*, 35 (1964).

Osborn, J. R., and Burick, R. J., "Technique for the Measurement of the Erosive Burning Rate of Solid Rocket Propellants." In *Nucleonics in Aerospace* (New York: Plenum Press, 1968), pp. 87-94.

Osborn J. R., Burick, R. J., and Panella, R. F. "Continuous Measurement of Solid Propellant Burning Rates," *Review of Scientific Instruments*, 37 (1966).

Osborn, J. R., Murphy, J. M., and Kershner, S. D., "Photographic Measurement of Burning Rates in Solid Propellant Rocket Motors." *Review of Scientific Instruments*, 34 (1963): pp. 305-306.

Zucrow, M. J., Osborn, J. R., and Murphy, J. M., "The Erosive Burning of a Non-Homogeneous Solid Propellant," 56th Annual AICHE Meeting (December 1963).

Zucrow, M. J., Osborn, J. R., and Murphy, J. M., "An Experimental Investigation of the Erosive Burning Characteristics of a Non-homogeneous Solid Propellant." *AIEE Journal*, 3 (1965): pp. 523-525.

Zucrow, M. J., Osborn, J. R., Murphy, J. M., and Kershner, S. D. "Final Report on the Investigation of Velocity Upon Burning Rate of Solid Propellants," Jet Propulsion Center, Purdue University, F-63-3 (December 1963).

Chapter 10

Fundamental Scientific Questions in the Early Period of Rocket Propulsion Development[1]

Martin Summerfield[2]

This memoir may be regarded as a sequel to the series of memoirs presented at previous Congresses of the IAF/IAA by Frank J. Malina, the American pioneer in rocketry who died during the past year. It is dedicated to his memory. The author had the good fortune to be his professional collaborator in rocketry during the early critical phase in the United States, as well as his personal friend. The author was also a close collaborator in this program and a friend of Theodore von Kármán, a founder of the International Academy of Astronautics and its first President, who led this early American research effort. Both Malina and von Kármán had important roles in shaping American rocketry and space exploration; and indeed, their roles acquired world significance as the years went by.

To those of us who were participants in the search for solutions to the many problems that arose in those years and who are therefore sensitive through experience to the various alternatives, it is immediately evident that today's rocket propulsion technology flows, in major ways, directly from the results of the work of the team that those two men formed and led. We have only to take a quick look at the types of liquid propellant rocket engines in use today—the injectors, the chamber configurations, the nozzle designs, the cooling techniques, etc.; and in the types of castable, composite solid propellant engines—the grain configurations, the igniters, etc.; to appreciate immediately the clear line from those early years.

---

[1] Presented at the Sixteenth History Symposium of the International Academy of Astronautics, Paris, France, 1982.

[2] Princeton Combustion Research Laboratories, Inc., Princeton, New Jersey, U.S.A.

In this paper, I will present some of the critical issues that arose in the very earliest stage of liquid propellant rocket engine development. Much of the wisdom that governed the final technical choices was due to von Kármán and Malina; the author was their associate and the performer of the investigations in the particular problems described here. In recounting this early history of the search for technological solutions, it is hoped that useful principles will become evident to the reader on how scientific minds approach and solve difficult engineering questions.

This history goes back to the 1940-1942 period at the California Institute of Technology, when the principal question facing Theodore von Kármán and his little group of post-docs was simply this: Is high performance rocket flight feasible, or are there fundamental thermodynamic, physical, and/or chemical kinetic barriers that stand in the way? More specifically:

1. Do chemical rocket propellants, even the most energetic combinations known to chemists, possess enough energy to permit, say, escape from the Earth? Is there a simple equation that might serve to reassure us, that the long-range objective is indeed possible, granted that some formidable engineering problems would have to be solved? No rocket had yet achieved significant flight performance.
2. Can the combustion reaction be carried out with high efficiency in a volume small enough to be acceptable in a weight-limited rocket vehicle, when the only practical, high-efficiency combustion chambers known to engineers at the time were a thousand-fold and more greater than anything that could fly? Were there fundamental chemical kinetics limitations that would prevent us from reaching the goal of a high performance rocket that could pull away from the Earth's gravitational field? Were the practical combustion engineers justified in their skepticism?
3. Can a long-duration rocket motor based on liquid propellants be regeneratively cooled so as to permit steady-state operation without burnout, and especially, would this be possible for the high temperature, efficient combustion that would be needed in order to make space flight a reality? If one believes that rocket motors are similar to conventional power plants, with their typical 25 percent to 50 percent heat rejection to the coolant, the whole idea of a regeneratively cooled rocket motor would be impossible. Was there a practical way around this seemingly difficult obstacle? No one had yet found one. Was there a fundamental reason?
4. Is stable combustion possible? True, instability had never posed an insurmountable barrier to man's development of novel systems; a means to stabilize any kind of novel system had always been found in the past. Still, the question had to be raised at the outset: would a basic source of instability be encountered that might not be solvable in a straightforward manner?
5. Can one find a practical self-pumping system that could generate the required propellant injection pressures on a sustained basis that would be light enough to fly in a high-performance rocket vehicle? Engine-driven pumping systems depend on air for combustion power, and a self-contained air supply would be far too heavy to carry along, as in the torpedoes of the time. Pressure-feed systems are also too heavy. No other practical systems were known at the time. Would this problem turn out to be the "Achilles' Heel" of the escape-from-the-Earth problem?

The Caltech group, led by the thinking of von Kármán, posed these basic questions early in the beginning of the rocket program. In fact, assailed by doubts on all these points, the attitude of von Kármán was to go slowly, at first, not to promise too much, to keep a "low profile" about the project, to hold spending at a very low level, to hold back on hiring so-called permanent staff, and generally to refrain from making a firm commitment. We must remember, in the year 1940, despite the previous 20-odd years of work by Robert H. Goddard the American rocket pioneer, everything that was known, either by practical test or theoretically, pointed toward negative answers to the above five questions. Could the work of Goddard have been misleading? Why, after two decades of work, had high performance rocket flight eluded him? Were the above problems indeed so fundamental as to make rocket flight improbable, or worse, impossible? If the barriers could be shown to be fundamental on theoretical grounds, then there would be no point in making all these efforts, and no point in taking the sponsor's funds, tiny though they were (less than $20,000 per year). The years 1940-1941 were the critical ones for von Kármán and Malina, while these questions were being analyzed and a decision made to go forward or not.

One by one, the above-listed questions were resolved, all of them in the affirmative, and the decision was reached by von Kármán that high performance rocketry would indeed be possible. It would now be the task of clever engineers to work out the practical solutions. How these problems were resolved, at first theoretically and then by confirming experiments, is explained elsewhere in scientific and engineering terms.

Although these questions were faced also by the German rocket engineers during the decade prior to 1942, none of their accomplishments were known in the United States until nearly the end of World War II (late 1943, to be exact). Our purpose in briefly discussing this early history of Caltech rocket work was to set forth the fundamental questions identified, the questions that had to be resolved in order to make high performance rocket flight possible, and to set forth the scientific logic of the attack on these questions as it was pursued by the von Kármán-Malina team.

Chapter 11

Scientific Foundations for the Implementation of Human Space Flight[1]

V. I. Yazdovsky and Y. Vedeshyne[2]

Scientific foundations for the possibility of human flight into space and its implementation have been worked out systematically by a large team of Soviet scientists over many years. The results of the first studies (1948-1952) showed that a sealed cabin of small volume with a life-support system used during a rocket flight up to the altitude of 100.8 km, created conditions necessary for animals to stay in space for up to three hours. The effect of flight factors (overloading, weightlessness, cosmic rays, etc.) did not practically cause changes in behavior and conditions of separate physiological functions in animals.

When carrying out the second stage of studies, it was made clear that space suits without masks provide necessary conditions for the animals' life during their flight in a nonsealed cabin of the rocket up to the altitude of 110 km. There were no noticeable changes in the condition of the separate physiological functions in the animals. The ejection at the altitude of 75-86 km with a speed of 565-728 m/sec (2030-2606 km/hour) and at the altitude of 39-40 km with a speed of 1020-1150 m/sec (3672-4160 km/hour) is a reliable method of separating the cabin from the rocket.

The animals' flight in a rocket up to the altitudes of 212 and 450 km did not reveal noticeable changes in behavior and conditions of physiological functions in the animals. The principal result of the biological experiment with the second man-made Earth satellite showed that the animals' flight in rockets with conditions nearer to those of space

[1] Presented at the Fifteenth History Symposium of the International Academy of Astronautics, Rome, Italy, 1981.

[2] National Committee for the History of Science and Technology, U.S.S.R. Academy of Sciences, Moscow, U.S.S.R.

flights are satisfactorily endured by the animals to an even greater degree. The biological tests, performed in a space flight of more than 25 hours with 17 full revolutions around the Earth, resulted in obtaining unique scientific data on the influence of factors of space on the cytological systems of living organisms. The implemented flight convinced Soviet scientists in the correctness of chosen trends in manned space flight training and outlined a concrete path to implement such flights.

After adoption of a scientific program for human flight of one-revolution's duration, a series of qualified flights of animals (Chernushka, Zvezdotchka, and other biological objects) on space-ships/sputniks for one revolution around the Earth was conducted. During space flight, all systems of life-support and individual safety facilities were maintained. At the same time, scientific data was accumulated on the influence of space flight on a whole series of biological objects which were in flight on the second Soviet space-ship/sputnik. After successful accomplishment of the necessary scientific research, which laid foundations for the possibility of manned space flight, and as a result of the testing of all space flight safety systems and determination of selection and training of cosmonauts, Soviet scientists prepared themselves for the implementation of the dream of mankind: human flight into space.

All this made it possible for the Soviet Union to carry out, for the first time in history, a space flight with a man aboard the spaceship Vostok in the spring of 1961.

Chapter 12

The History of Extravehicular Activity (EVA) in U.S. Human Spaceflight[1]

R. Scott Millican[2]

The history of extravehicular activity (EVA) in U.S. human space flight goes back to the early 1960s. It has become a major part of American activities, and perhaps Lunar surface exploration would not have been practical without the astronaut's capability to leave the spacecraft to retrieve lunar samples, deploy experiments, and make on-the-spot observations, judgments, and decisions relative to acquiring scientific data. EVA capability was further demonstrated during the Skylab Program when several major space station repairs were accomplished by EVA astronauts. These repairs allowed mission continuation and were considered the major factor in successful completion of the Skylab program objectives. EVA is now such a vital and integral part of U.S. manned space flight that most advanced spacecraft and space station studies include EVA capability for routine maintenance, construction, and servicing tasks.

Requirements for EVA

Although early requirements for EVA seemed to be associated with expanding humanity's general capability to work in space, the primary objective was to develop and demonstrate EVA systems and operations for eventual lunar exploration. In 1959, NASA's Ten Year Plan projected manned lunar flights no earlier than the 1970s. On 25

[1] Presented at the Fifteenth History Symposium of the International Academy of Astronautics, Rome, Italy, 1981.

[2] Scott Science and Technology, Houston, Texas, U.S.A.

May 1961, however, President John F. Kennedy committed the United States to landing an American on the Moon before the end of the decade. This objective resulted in the requirement for a much earlier development of EVA capability than had been expected.

Project Mercury was just beginning as Alan B. Shepard made the first suborbital flight on 6 May 1961. A "Mark II" version of the Mercury capsule came under immediate development to bridge the gap between the one-person capsule and a three-person lunar spacecraft, already on the drawing boards. This two-person spacecraft was required to support the development of rendezvous and docking, in addition to other requirements, such as EVA.

Early Leaders in EVA Development

In 1958 NASA established a Space Task Group at Langley Laboratory, Hampton Roads, Virginia, to put an American in space. Robert Gilruth, who had spent years as an engineer with NASA's immediate predecessor, the National Advisory Committee for Aeronautics, headed this organization, and by mid-1959 three divisions had been formalized. Maxime Faget was leader of the Flight Systems Division, Charles Mathews managed the Operations Division, and James Chamberlin led the Engineering and Contracts Division. Although primarily charged with developing the Mercury Program, in 1961 this group moved into what was later called Project Gemini, when it was determined that an improved Mercury capsule was needed.

Chamberlin took the early lead in managing the upgraded design effort which the contractor, McDonnell Aircraft, was supporting. The Mercury spacecraft required new operational features to support EVA, such as more volume, an egress hatch, and a second pilot capability. Since it was apparent that the Mercury had to be redesigned, Max Faget requested that John Yardley, McDonnell's manager for Mercury operations at Cape Canaveral, look into the possibility of a two-person spacecraft for the purpose of supporting EVA, in addition to other reasons. This occurred in March 1961. The requirement became almost imperative when the President made his Moon landing commitment in May. Chamberlin then pursued the Mercury Mark II two-person version with considerable effort, and NASA's head of Space Flight Programs, Abe Silvertein, endorsed the design by mid-year.

Several years passed as the Mercury Program was completed and two unmanned Gemini flights were conducted before serious EVA talks arose. The Soviets had performed the first EVA during the *Voskhod 2* mission on 18 March 1965. With the first Gemini manned flight scheduled for 23 March of the same year, the Americans would have to settle for an EVA on *Gemini 4*, scheduled for June 1965, to stay in the "space race." Had it not been for *Gemini 4* astronauts pressing for early delivery of EVA spacesuits, the opportunity for EVA would have been delayed until later in the Gemini program. Gilruth, by then director of the newly created Manned Spacecraft Center at Houston, Texas, convinced NASA Administrator, James Webb, that the crew was trained and all EVA hardware was qualified, including a hand-held maneuvering unit (HHMU). On 25 May 1965, the press was notified that Edward H. White II would become the first American astronaut to walk in space.

The Gemini Experience

EVA was performed on five of the ten piloted Gemini missions. The apparent successful EVA by Ed White on *Gemini 4* was quite dramatic as colorful EVA photos were returned from this flight (Figure 1). For one-half hour, White floated around with no disorientation attached to a life-support umbilical, and he successfully operated an HHMU. Two problems arose in this first experience with EVA. First, the heat removal capability of the EVA life-support unit was insufficient to handle the body heat generated by the astronaut's exertion, and White became too hot inside the suit. This had to be dealt with in a redesign of the spacesuit. The astronauts also had difficulty closing the hatch. This was an especially important problem because if the hatch could not be resecured, the astronauts could not return safely to Earth. They jostled and muscled the hatch closed, and it, too, had to be modified for future flights. The next EVA, scheduled for *Gemini 8*, was to evaluate a new life-support system and a backpack mounted Astronaut Maneuvering Unit. Due to a spacecraft problem resulting in early mission termination, however, Dave Scott was unable to perform the EVA.

Figure 1 Edward H. White II is shown performing his *Gemini 4* EVA during the third orbit. White floats into space, secured to the 25-ft umbilical line and a 23-ft tether line, both wrapped together with gold tape to form one cord. White became the first American astronaut to egress his spacecraft while in orbit. He remained outside the spacecraft for a total of 21 minutes. White wears a specially designed space suit for his EVA. He is holding a Hand-Held Self-Maneuvering Unit which he used to move about in the weightless environment. White and the GT-4 command pilot, James A. McDivitt, performed other scientific and engineering experiments before completing their 62 revolution mission and returning safely to Earth (NASA Photo No. 65-H-1024).

The EVAs that took place on the flights of *Gemini 9* and *11* saw a replay of basically the same problems as White had encountered on *Gemini 4*. Both Gene Cernan and Dick Gordon became overworked due to inadequate heat venting in the suits. The life-support system was not capable of removing the high heat loads as helmets fogged and sweat ran into their eyes causing early EVA termination. *Gemini 10*, however, flown between these missions, produced more favorable results. Mike Collins successfully used an HHMU and retrieved an experiment from a docking target vehicle. The EVA was planned for only one daylight pass but was shortened due to a spacecraft problem to only 39 minutes.

In spite of these earlier efforts, prior to *Gemini 12* NASA still had not successfully demonstrated the operational capability of EVA to satisfy the lunar landing commitment. The training for the final EVA concentrated on short tasks with rest breaks, improved foot restraints, and brought about EVA end-to-end underwater crew training. These features, along with more disciplined, well-organized EVA pre-egress procedures, led to a totally successful *Gemini 12* EVA performed by Buzz Aldrin.

Adequate simulation of the EVA environment, proper foot restraints, and sizing the life-support system for greater heat dissipation were major lessons learned during Gemini. By the conclusion of the Gemini program EVA was considered an operational capability of U.S. human spaceflight.

Apollo

Of the four piloted Apollo test flights, only one EVA evaluation was conducted. With the Lunar Module and Command Module docked, *Apollo 9* astronauts depressurized both spacecraft to demonstrate a contingency EVA transfer procedure in the event of a blocked tunnel passageway. Astronaut Rusty Scheickart, wearing a new spacesuit and portable life-support system, egressed the Lunar Module and, using foot restraints and hand rails developed on Gemini, successfully moved to the Command Module. Dave Scott waited in the opened Command Module hatch during his standup EVA.

All lunar surface EVAs were very successful. *Apollo 11*, *12*, and *14* backpack life-support systems, attached to the A7L spacesuit, provided four to five hours EVA time (Figure 2). *Apollo 15*, *16*, and *17* missions utilized a life-support system providing seven to eight hours EVA time. An improved spacesuit, the A7LB, also allowed more waist mobility, making it possible for astronauts to ride the Lunar Rover in a sitting position during these missions (Figure 3).

The last three flights also required EVA during the return to Earth phase, in order to retrieve film from a lunar surface mapping camera located in the service module (Figure 4). An umbilical cord attached to the spacecraft was used for life-support.

Zero-gravity EVA training was conducted using, primarily, underwater facilities for simulation, while lunar surface training was effectively accomplished by one-gravity walk-throughs. The Apollo Program was designed specifically with EVA as a primary requirement for mission success. All aspects of EVA operations performed well, including life-support equipment, procedures, real-time systems monitoring inflight and on the ground, EVA support equipment, and training.

Figure 2 *Apollo 11* EVA. Neil A. Armstrong, Michael Collins and Buzz Aldrin were launched to the Moon by a Saturn V launch vehicle 9:32 a.m. EDT July 16, 1969 from Complex 39A, Cape Kennedy, Florida. Armstrong and Aldrin landed on the Moon on July 20, 1969 and, after take-off from the Moon on July 21, joined Collins in the Command Module circling the Moon. The astronauts splashed down in the Pacific Ocean and recovery was made by the U.S.S. *Hornet* at 12:50 p.m. EDT July 24, 1969 (NASA Photo Nos. 69-HC-695, 69-H-1266, AS11-40-5872).

Figure 3 Apollo Lunar Rover EVA. Astronaut Irwin is shown walking away from the Rover which is parked near the edge of Hadley Rille, far wall of the rille is in the distance at extreme upper left. *Apollo 15* was launched July 26, 1971 at 9:34 a.m. EDT and touched down at the Hadley-Apennine site at 6:16 p.m. EDT, July 30—staying a total time of 66 hours and 55 minutes. Splashdown in the Pacific Ocean was on August 7 at 4:46 p.m. EDT. Astronaut Alfred Worden was the command module pilot, James Irwin, lunar module pilot and David Scott, commander (NASA Photo Nos. 71-H-1418, AS15-82-11168).

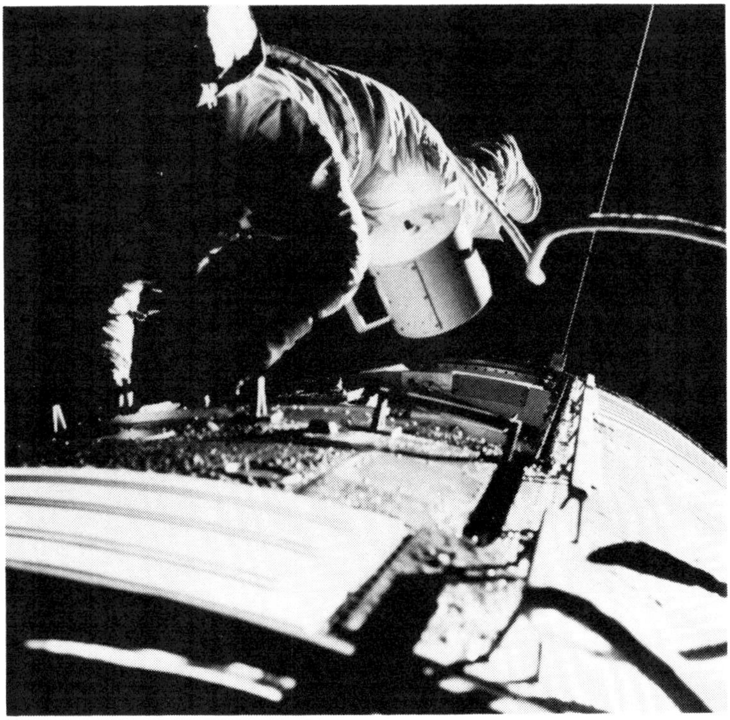

Figure 4 *Apollo 17* astronaut Ronald E. Evans as he performs an EVA—retrieving film canisters from the Lunar Mapping Camera and Panoramic Camera during the *Apollo 17* spacecraft's trans-Earth coast. The cylindrical object at Evans' left side is the Mapping Camera cassette. The total time for the trans-Earth EVA was one hour seven minutes 18 seconds, starting at ground elapsed time of 257:25 (2:28 p.m.) and ending at ground elapsed time of 258:42 (3:35 p.m.) on Sunday, December 17, 1972 (NASA Photo Nos. 72-HC-925, 72-H-1574, AS17-152-23391).

Skylab

For the Skylab Orbital Laboratory, EVA became not just a planned mode for acquiring scientific data but also a critical skill in activating the spacecraft after damage during launch. Planning for Skylab EVA began shortly after completion of the Gemini Program; it was emphasized as a technique for retrieving and replacing film from solar observation cameras. Four complete changeouts of film canisters were to be included in six 2-1/2 EVA periods.

In addition to the successful film operations, major repair activities were undertaken on each of the three Skylab missions. A total of 10 EVAs were actually performed, for a total of over 82 hours, two-thirds more than planned. Eighteen new mission objectives and 13 inflight repair tasks were added to the originally planned EVA requirements, and all were accomplished. Three significant EVA repairs prevented mission termination; these included manual deployment of a solar array panel, covering a portion of the Orbital Workshop with a reflective heat shield, and replacement of a rate gyro processor (Figure 5).

Figure 5 Two astronauts selected for the Skylab mission are assisted by scuba divers during an Neutral Buoyancy Simulator (NBS) test for Skylab EVA training. During the three manned missions of the Skylab program, a number of EVAs were necessary to install and retrieve ATM film and experiment D024. The NBS at Marshall Space Flight Center had an important role during the design phase of the Skylab systems, and is presently being used effectively to simulate EVA maneuvers. The purpose of underwater Skylab EVA simulation was twofold: (1) to provide time-motion data for updating and finalizing the crew EVA checklists; (2) to train the crew for the actual EVAs by running them fully pressure-suited through simulated zero-g EVA maneuvers on a second-by-second basis using preliminary and actual flight check-lists (NASA Photo Nos. 72-H-1093, 72-HC-607).

Maximum use of humanity's adaptability in space was demonstrated by including basic EVA provisions. Since all Skylab astronauts had received extensive EVA underwater training for planned film changeout tasks, new real-time EVA requirements presented no major difficulties, if adequate translation aids and foot restraints could be provided. All new tasks were evaluated in a large water tank, which contained a full scale model of the Skylab Workshop where EVA operations were performed. Common hand tools and special application tools were frequently used.

The life-support system was a new design and relied on a 60-foot umbilical to supply needed consumables from the spacecraft to the belly mounted pressure control unit which was attached to the A7LB spacesuit. A secondary oxygen pack provided a 30-minute emergency reserve and was attached to the leg.

Space Shuttle

EVA during the Space Shuttle era may become as common as space flight itself. Every flight will have two qualified EVA crewmen. Accommodations include a new spacesuit and an integral backpack life-support system capable of fitting a wide range of male and female crew members. Flood lights and a TV camera are built into the helmet (Figure 6). Also available are standard EVA hand tools, portable foot restraints, and a Manned Maneuvering Unit. The MMU is a propulsive backpack device which will allow the crewman to translate to areas beyond the cargo bay. It has six-degree-of-freedom control authority and an automatic attitude-hold capability. Range will be up to 100 meters or more.

Figure 6 Space Shuttle EVA. Bruce McCandless II, on February 7, 1984 used the combination of the Remote Manipulator System (RMS) arm and the Mobile Foot Restraint (MFR) to experiment with a "Cherry-Picker" concept (NASA Photo Nos. 84-HC-96, 84-H-93, 584-27039).

The astronauts on the first Shuttle flight were trained for several potential contingency operations. Simple EVA features have been designed into the payload bay doors to allow manual, backup door closing by an EVA crewman. A thermal tile EVA repair capability was under development for the first flight but was not carried. It included a uniquely designed work restraint unit that could be secured to the spacecraft exterior by a crewman flying the MMU (Figure 7). Carrying tools, repair materials, and foot restraints, it could be moved from one location to another.

Various studies have shown that use of EVA in payload systems design can enhance operational success for tasks such as deployment, retrieval, and maintenance, as well as for correcting malfunctions. Experienced EVA operations personnel participating in early STS payload design activities can assist in achieving simple, cost effective operational concepts.

Current studies being conducted by NASA on permanent space stations include EVA for construction of large structures and for satellite servicing. History has shown that maximum space flight operational flexibility is achieved when EVA capability is provided.

Figure 7 Space Shuttle Manned Maneuvering Unit. On February 8, 1984, Bruce McCandless II, one of two 41-B mission specialists, participated in an historical EVA. He is shown a few meters away from the cabin of the Earth-orbiting Space Shuttle *Challenger* in this 70 mm frame. This spacewalk represented the first use of a nitrogen-propelled, hand-controlled device called the Manned Maneuvering Unit (MMU), which allows for much greater mobility than that afforded previous space-walkers who had to use restrictive tethers. Robert L. Stewart, tested another similar unit two days later. Inside the spacecraft were Astronauts Vance D. Brand, commander; Robert L. Gibson pilot; and Ronald E. McNair, mission specialist (NASA Photo Nos. 84-HC-74, 84-H-71).

Table 1
EVA EXPERIENCE SUMMARY

MISSION DATE	EVA NO./CREWMAN	EVA TIME TYPE	TIME	CUM	OBJECTIVE	REMARKS
Gemini IV June 1965	1/White	UMB	:36	:36	Prove EVA feasibility; maneuver with Hand Held Maneuvering Unit (HHMU)	EVA worked as planned, including HHMU.
Gemini VIII March 1966	1/Scott	-	-	- :36	Evaluate new self-contained life support system and freefly HHMU.	Mission aborted prior to EVA.
Gemini IX June 1966	1/Cernan	UMB	2:07	2:43	Evaluate new life support system; fly new Astronaut Maneuvering Unit; retrieve experiment	High workload and helmet fogging caused early termination. Maneuvering Unit not used.
Gemini X July 1966	1/Collins 2/Collins	STUP UMB	:50 :39	4:12	EVA familiarzation; retrieve experiment evaluate HHMU, photography.	First transfer of tethered crewman between undocked vehicles.
Gemini XI September 1966	1/Gordon 2/Gordon	STUP UMB	2:10 :33	6:55	Perform simple work tasks, evaluate HHMU, photography.	High workload led to early EVA termination. Standup EVA provided familiarization.
Gemini XII Nov. 1966	1/Aldrin 2/Aldrin	STUP UMB	3:24 2:06	12:25	Evaluate matrix of simple tasks, evaluate translation and restraint aids, photography.	Underwater training led to first tangible results of EVA capabilities.
Apollo 9 March 1969	1/Schweickart Scott	Free STUP	:47 :47	13:59	Demonstrate LM to CM transfer capability, demonstrate adequacy of Apollo EVA equipment and procedures.	All objectives satisfied.

Table 1

EVA EXPERIENCE SUMMARY (CONTINUED)

MISSION DATE	EVA NO./CREWMAN	EVA TIME TYPE	TIME	CUM	OBJECTIVE	REMARKS
Apollo 11 July 1969	1/ Armstrong Aldrin	Free Free	2:48 2:48	19:35	Demonstrate lunar surface EVA capability, gather lunar samples.	All objectives satisfied.
Apollo 12 Nov. 1969	1/Conrad Bean 2/same	Free Free Free	4:00 4:00 3:46 3:46	35:07	Deploy experiment hardware, retrieve lunar samles, retrieve samples from Surveyor spacecraft.	All objectives satisfied.
Apollo 13 April 1970	1/Lovell Haise 2/same	- -	- -	35:07	Deploy scientific equipment, collect lunar samples.	Mission aborted prior to EVA.
Apollo 14 Jan. 1971	1/ Shephard Mitchell 2/same	Free Free Free	4:48 4:48 4:35 4:35	53:53	Deploy scientific equipment, collect lunar samples.	All objectives satisfied.
Apollo 15 July 1971	1/Scott 2/Scott Irwin 3/same 4/same 5/Worden Irwin	STUP Free Free Free Free UMB STUP	:33 6:33 6:33 7:12 7:12 4:50 4:50 :38 :38	92:52	Deploy scientific equipment, collect lunar samples, retrieve film from CSM SIM Bay during transearth coast.	All objectives satisfied. Modified LM and EVA equipment allowed longer stay time and EVA time. First use of lunar rover. Only standup EVA from LM during Apollo.
Apollo 16 April 1972	1/Young Duke 2/same 3/same 4/ Mattingly Duke	Free Free Free Free UMB STUP	7:11 7:11 7:23 7:23 5:40 5:40 1:23 1:23	136:06	Deploy scientific equipment, collect lunar samples, retrieve film from CSM SIM Bay.	All objectives Satisfied.
Apollo 17 Dec. 1972	1/Cernan Schmitt 2/same 3/same 4/Evans Schmitt	Free Free Free Free UMB STUP	7:12 7:12 7:37 7:37 7:15 7:15 1:06 1:06	182:26	Deploy scientific equipment, collect lunar samples, retrieve film from CSM SIM Bay.	All objectives satisfied.

Table 1
EVA EXPERIENCE SUMMARY (CONTINUED)

MISSION DATE	EVA NO. / CREWMAN	EVA TIME TYPE	TIME	CUM	OBJECTIVES	REMARKS
Skylab 2 May 1973	1/Weitz	STUP	:35		Attempt to free solar array system for full deployment.	Task unsuccessful during standup EVA from CM.
	2/Conrad Kerwin	UMB	1:36 1:36		Deploy OWS solar array system.	Task successful most significant contingency EVA to date.
	3/same	UMB UMB	3:23 3:23	192:59	Retrieve and install film cannisters.	Film changeout successful. Also repair one camera and an electrical current regulator.
Skylab 3 July 1973	1/Lousma Garriott	UMB UMB	6:31 6:31		Deploy twin pole sunshade (contingency operation), install film cannisters. Repair camera. Deploy experiment.	All tasks successful.
	2/same	UMB UMB	4:30 4:30		Replace film, repair two cameras and rate gyroscopes.	All tasks successful.
	3/Beam Garriott	UMB UMB	2:41 2:41	220:23	Retrieve film, repair one camera, retrieve samples.	All tasks successful.
Skylab 4 Nov. 1973	1/Gibson Pogue	UMB UMB	6:34 6:34		Install film, repair one camera, operate experiment, install another experiment, deploy samples, repair experiment pointing antenna.	All tasks successful.
	2/Carr Pogue	UMB UMB	6:53 6:53		Replace film, retrieve experiment repair two cameras, operate several experiments.	All tasks successful.
	3/Carr Gibson	UMB UMB	3:29 3:29		Retrieve samples, operate several experiments, measure experiment temperature.	All tasks successful.
	4/Carr Gibson	UMB UMB	5:19 5:19	264:53	Retrieve film, operate experiment, retrieve experiments	All tasks successful.

Bibliography

Bland, Daniel A., *Space Shuttle EVA Opportunities*. Houston, TX: Johnson Space Center, JSC-11391, 1976.

Covington, J. H., and Millican R. S., *Apollo EVA Postflight Report*, Internal Johnson Space Center Memorandums, 1962-1972, Houston, TX.

Hacker, Barton C., and Greenwood, James M., *On the Shoulders of Titans: A History of the Gemini Project*. Washington, D.C.: National Aeronautics and Space Administration, 1977.

Machell, R. M., editor. *Summary of Gemini Extravehicular Activity*. Washington, D.C.: National Aeronautics and Space Administration, NASA SP-149, 1967.

Schultz, D. C., *Extravehicular Activity Experience in Manned Space Operations*, NASA General Working Paper, MSC-05878, 1970.

Schultz, D. C., Kain, R. R., and Millican, R. S., "Skylab Extravehicular Activity," *Advances in the Astronautical Sciences* (The Skylab Results), Vol. 31, Part I, American Astronautical Society, AAS 74-120, 1975.

Chapter 13

Space Research in Poland after 1958[1]

M. Subotowicz[2]

The Space Research Committee (SRC) of the Polish Academy of Sciences (PAS) has guided the Polish activities in space research since 1964. Even so, the first network of stations observing satellites was established in Poland beginning in 1958 in response the Sputnik launches. From 1958 until 1973 the SRC worked closely with a group of young rocket constructors in Cracow that made meteorological rocket experiments. During this period Poland conducted, in 1967, the first common space experiment in cooperation with the Soviet Union. Since 1973, however, the meteorological observations exploiting satellite data have been carried out by the state institute for Meteorology and Water Supply. Beginning in 1976, Poland has also belonged to the international space agency "Intercosmos," and this group has been especially important in charting the present direction of Polish space activities. It was this group that was instrumental in chartering in Poland the Space Research Center of PAS, and in 1977 started the program "Interplanetary and near-Earth space research and its practical use." In the framework of the Intercosmos-Organization, Poland has taken part in the following investigations: space physics, meteorology, space communication, space biology and space medicine, manned space flights, remote sensing and space processing (material sciences). Rockets and satellites are supplied by the Soviet Union.

The Polish program of space physics investigation has centered on research concerning the upper atmosphere and ionosphere, the influence of solar activity, and on the processes of plasma in interplanetary space around the Earth. There are in preparation spectrometric and photometric measurements in X- and UV-ray region from onboard solar satellites. The experiment also investigated the interaction of primary cosmic radia-

[1] Presented at the Fifteenth History Symposium of the International Academy of Astronautics, Rome, Italy, 1981.

[2] Institute of Physics, M. Curie-Skodowska University, Lublin, Poland.

tion with the nuclei of photographic emulsion. Theoretical analyses of different photochemical models of the ionosphere and some problems of solar physics have also been conducted. Some experiments in space processing prepared at the Institute of Physics of PAS were also carried out in the spacecraft *Salyut 6* in July 1978 by the Polish astronaut, Col. Mirosaw Hermaszewski.

Satellite observations have been used for geodynamic research and developing observation techniques and methods of analysis. These established the methodology of using satellite observation for purposes of the basic geodetic network in the country. Satellite observations also helped elaborate the methods of the satellite cloud pictures analysis for meteorological purposes. Satellite measurements also have provided mean temperature profiles of the atmosphere for several locales through calculation of seasonal temperature variations. Different cycle changes of the atmosphere and their influence on the climate were analyzed.

Part V

Pioneers of Rocketry and Astronautics

Chapter 14

Konstantin Eduardovich Tsiolkovsky and the Present Times[1]

A. A. Kosmodemjansky[2]

The present brief paper is concerned with the analysis of the forecasts drawn from the fundamental discoveries and designs made by Konstantin Eduardovich Tsiolkovsky—a leading scientist and pioneer of spaceflight from Russia (Figure 1).

K. E. Tsiolkovsky was born on 17 September 1857 to the family of a forester in a village called Izhevskoye, in the district of Spassk, Rjazan province. Recalling his childhood, Tsiolkovsky wrote in his autobiography: "I loved reading passionately, and I read whatever books were available . . . I indeed liked to immerse myself in dreams . . . In my dreams I was taking high leaps, climbing up poles or ropes like a cat. Absolute lack of gravity was my dream."[3]

When ten years of age he was taken ill with scarlet fever. The illness became serious, and he nearly lost all of his hearing. "Because of deafness," he recalled, "my biography is but of low interest, since I was deprived of personal contacts, and hence of observations and adoptions. My biography is poor in personalities or personal intercourse."[4] From age 10 to 14 Tsiolkovsky reported that his experience was "of the most melancholy mood, being the darkest period of my life. . . . Nothing to speak of well during that time."[5] When about 14 years of age, while borrowing books from his fa-

---

[1] Presented at the Sixteenth History Symposium of the International Academy of Astronautics, Paris, France, 1982.

[2] U.S.S.R. Academy of Sciences, Moscow, U.S.S.R.

[3] K. E. Tsiolkovsky, "Features from my life," *Collections of K. E. Tsiolkovsky*, (Moscow: Aeroflot, 1939), p. 19.

[4] *Ibid.*, p. 20.

ther's library, he came across a work on arithmetic, and started studying it. It came easily to him, and he immersed himself in the reading of other books concerned with natural and physico-mathematical sciences. He recalled:

> I was enthralled with an astrolabe, taking a great interest in measuring distances to inaccessible objects, in taking plans, finding heights. And I had set up an astrolabe-goniometer. With the aid of this instrument, without leaving home, I had identified the distance to a fire-tower. I found it to be 400 arshins. Then I go out and check. Proved correct. From that time on I put trust in theoretical approach.[6]

Figure 1 Konstantin Eduardovich Tsiolkovsky (1857-1935), father of Russian cosmonautics and space philosopher. Courtesy of NASA History Office.

---

[5] *Ibid.*, p. 23.

[6] *Ibid.*, p. 24.

The inquiring mind and gift for original reflection manifested by young Tsiolkovsky were noted in the family, and his father urged him to move to Moscow, where the young man had made up his mind to study by himself. For three years, while working in libraries (from 16 to 19 years of age) he thoroughly studied courses in mathematics and physics. When 17 years old, he already "had perused courses in higher algebra, differential and integral calculus, analytical geometry, spherical trigonometry and the like."[7] At the same time he was preoccupied with varied scientific and engineering problems, and he did his best to resolve these with the aid of the knowledge acquired.

It was during these years that an idea arose in his mind as to how a person might fly within the atmosphere of the terrestrial globe as well as in outer space, by utilizing the properties of centrifugal forces. A machine, as conceived by Tsiolkovsky, would be comprised of a box (a cabin) wherein two pendulums with massive balls at the end of the rods were to execute an oscillating motion, while being mounted vertically upward, with the balls at the top. According to the concept developed, centrifugal forces produced by the oscillating rods would lift the cabin, carrying it over boundless distances into outer space. "Having conceived such a machine, I went into raptures over my invention," he remarked:

> and being unable to keep my seat I went outdoors to vent off the joy which could nearly choke me. An hour or two I was rambling about the streets of Moscow at night, meditating and verifying my discovery. Alas . . . The delusion vanished yet in the streets. However, the burst of ecstasy was so stunning, that throughout my whole life I saw this machine in my dreams: it was indeed an extraordinary fascination to levitate with the help of this craft.[8]

It is worth noting here that this faulty idea set forth by young Tsiolkovsky as to how to give motion to a flying vehicle through the action of inner (with respect to a given mechanical system) forces, cannot escape the minds of some contemporary inventors up to the present day. "Further fuel to the already blazing fire" had been added by some Russian popular scientific periodicals, which published inaccurate information on the invention made by an American. Suggestions had been made to exert motion to the center of mass of a mechanical system with the aid of a "rectifier of centrifugal forces" (a copy of Tsiolkovsky's idea), and to develop a range of apparatuses ("Inertiacidos") which were supposed to refute by way of experiment the law of conservation of momentum of a mechanical system without any external forces. In fairness, it took Tsiolkovsky but a single night to realize the falsity of his suggestion, while years are sometimes not enough for our contemporary inventors to give a fair analysis to their errors.

---

[7] *Ibid.*, p. 25.

[8] *Ibid.*, p. 26.

To follow a poorly grounded tradition in dwelling upon the incentives from which Tsiolkovsky's interest arose, in developing a theory of jet propulsion, one should mention the writings of Jules Verne and A. Fodorov, however, there are evidences that rocket dynamics and astronautics were dreamed about by Tsiolkovsky as early as his teen years.

He wrote in the "Introduction" to his *Exploration of Outer Space by Means of Jet Apparatuses* (1911-1912):

> It seems to me—wrongly, I believe—that the basic ideas and an unceasing flame rushing there—towards the Sun, releasing from the bonds of gravity, were within me from the very birthday. At any rate, I remember quite clearly that in early childhood, prior to books, my favorite dream was to dimly realize a medium free-of-gravity, where one can freely move towards any direction, feeling himself better than a bird in the sky.
>
> Up till now I cannot comprehend where these desires came to me from; there are no such fairy-tales either, but I had a dim faith, a sensation of and a desire for this very free-of-gravity medium. An old sheet of paper in my manuscripts comprising final formulae relating to a jet apparatus is dated as of August 25, 1898. Clearly, I had been previously concerned with such an apparatus.[9]

We are in a position to prove from archives and published papers that it was as early as 26 years of age when Tsiolkovsky realized that a rocket is an apparatus with the aid of which mankind "would timidly penetrate beyond the atmosphere, subsequently conquering for his own sake the entire circumsolar space." In 1954, the Soviet Union published a scientific diary compiled by Konstantin Eduardovich Tsiolkovsky, titled *Free Space*, all the entries thereto were made in the year of 1883. This diary includes considerations given to regularities inherent to the simplest phenomena of mechanical motion which is not affected by gravity or forces of resistance produced by a medium (aerodynamic forces). The basic conclusion arrived at by Tsiolkovsky is as follows: within a free space, the most rational way of imparting motion to some stationary object ("body"), or of changing motion being executed by a body is to throw off particles of a matter, i.e., with the aid of a reactive force. Should one be able to direct, at will, a jet along any of the three mutually normal directions, one would be capable of ensuring any curvilinear motion of the body.[10] Incidentally, as early as in 1883, Tsiolkovsky clearly realized that the simplest way of setting or changing motion within a free space was to utilize the reaction of an out-flowing gas jet.

In his first publication concerned with the theory of jet propulsion, titled *Exploration of the Outer Space by Means of Jet Apparatuses* (1903),[11] Tsiolkovsky derived a number of fundamental formulas, which enable one easily to mathematically establish the flight performance of rockets. Tsiolkovsky's formula, which gives the maximum

---

[9] K. E. Tsiolkovsky *Selected Works* (Moscow: U.S.S.R. Academy of Science, 1962), p. 167.

[10] *Ibid.*, pp. 11-12.

[11] *Ibid.*, pp. 136-66.

capabilities of the jet-propulsion technique of imparting motion within a free space, has enjoyed world-wide fame.[12] Using the up-to-date designations, this formula can be written as follows:

$$\vartheta_{max} = V_r \, ln\,(1 + Z) \cong 2.3 \; V_r \, lg\,(1 + Z) \qquad (1)$$

where ϑ_{max} is the maximum velocity of the rocket; V_r is a constant relative velocity of exhausted particles; $Z = m/M_k$ is Tsiolkovsky number (m = mass of fuel within the rocket; M_k = mass of the rocket without fuel). If for example, $Z = 9$, and $V_r = 3440$ mps, then $\vartheta_{max} = 7912$ mps, i.e., equal to the orbital velocity as applied to the planet Earth.

A critical part of Tsiolkovsky's work was a 1903 essay, titled "Among gravity: Vertical return to the Earth," where he first calculated formulas for establishing a storage of propellant required for a "soft" landing onto a planet without atmosphere.[13] He developed a relatively simple solution to the problem of establishing a propellant tankage capacity for a rocket which must acquire velocity ϑ_1, within a free space, and then suppress it. The formula is as follows:

$$\frac{M_{init}}{M_{fin}} = (1 + Z_1)^2 - 1 \qquad (2)$$

where Z_1 is Tsiolkovsky number, which ensure the attainment of velocity ϑ_1.

If, as is the case for the previous example, $\vartheta_1 = 7912$ ($Z_1 = 9$), then $M_{init}/M_{fin} = 100 - 1 = 99$. This means that to reduce soft landing into practice calls for a rather large storage of propellant, being as yet impractical as applied to a single-stage rocket. Note, that unfortunately it is rather seldom that a rudimentarily simple formula (2) given by Tsiolkovsky is mentioned in our home literature on rocket dynamics.

If we assume that the field of gravity of a planet (for example, Earth) is uniform with a certain range of altitudes, and that free fall acceleration is constant, being equal to $g = 9.81$ m/s^2, then, provided that acceleration accounted for by a reactive force is also constant,[14] being equal to p, we set:

$$\frac{M_{init}}{M_{fin}} = \left(1 + \frac{pZ_1}{p - g}\right)^2 - 1 \qquad (3)$$

---

[12] It will be noted here that in his science-fiction story "Beyond the Earth" (the first part of which was written in 1896) K. E. Tsiolkovsky considered motion of a rocket within a uniform gravity field, presuming that at the ascending powered rectilinear section of the flight, acceleration of the center of mass of a rocket will be at a constant level. In this case, equation for the motion of the rocket will be $Ma = \phi - Mg$, where a = const, $\phi = -dM/dt\, V_r$ – reactive force, g = const, is the force of gravity. Thus $dM/M = -[(a + g)/V_r]dt = -adt$ and $M = M_o e^{-\alpha t}$, where α = const., i.e., in this case, mass of the rocket will be changed according to the potential law. This case had been analyzed by K. E. Tsiolkovsky in great detail in his work published in 1903.

[13] Tsiolkovsky, *Selected Works*, pp. 158-161.

[14] In this case, the reactive force must be proportional to the mass of a rocket, and hence, the mass of the rocket would be changed according to the exponential law.

If $p = 4g$, and $z_1 = 9$, then $M_{init}/M_{fin} = 168$, which conclusively shows how difficult it is to fulfill a soft landing onto a free-of-atmosphere planet which has a gravity field similar to that of the Earth.

In 1929, Tsiolkovsky, taking into account the practically attainable velocities of particles exhausted from a nozzle of a jet engine (i.e., real values for v_r), developed a theory for multistage rockets (or "rocket trains" as he called them), having proved through a strict mathematical way that even with practically attainable figures for z one can achieve fairly high ratios between the initial mass of a rocket and its final mass.[15] We will point out here that the U.S. Apollo used this case in performing a soft landing onto the surface of the Moon, followed by successful return to the Earth:

$$\frac{M_{init}}{M_{fin}} = 3140/5 = 628$$

We would also like to point out here that the upper stages of the modern large ballistic types of rockets can be converted to winged flying vehicles that can then be employed in making intercontinental passenger flights in minimal time. From the theory of elliptical trajectories, the flight time could be established (approximately, taking no account of the atmosphere) as given in Table 1 below.[16]

Table 1

Distance of Flight (Kilometers)	Maximum Time of Flight
5,566	22 min 06 sec
10,019	32 min 14 sec
15,585	40 min 13 sec
20,038	42 min 13 sec

As can be seen from the formula given by Tsiolkovsky, the increase in the maximum velocity of a rocket can be achieved by means of increasing either the relative velocity v_r of particles being exhausted, or Tsiolkovsky number z. One can immediately appreciate that to improve v_r calls for advanced design of jet engines and for higher calorific power of the propellant components (combustible fuel and oxidizer).

Today, in modern U.S. laboratories, scientists have tested many promising fuel pairs for liquid-propellant rocket engines as:

1. Liquid fluorine + liquid hydrogen ($v_r = 4500 - 4700$ m/s is achieved);

---

[15] Tsiolkovsky, *Selected Works*, pp. 367-393.

[16] A brief description of the optimum elliptical trajectories can be found in A. A. Kosmodemjansky, *Course of Theoretical Mechanics* (Moscow: U.S.S.R. Academy of Sciences, 1975), Part 1, pp. 246-265.

2. Combination propellant: liquid fluorine + liquid hydrogen + lithium (v_r up to 6000 m/s is achieved).

Rocket engineers all over the world are dreaming of mastering monatomic hydrogen (estimated exhaust velocity v_r is up to 10,000 m/s), as well as of developing and mastering gaseous-phase reactors with the process medium (hydrogen), whereby an estimated v_r = 20,000 to 25,000 m/s can be achieved.

Simple calculations show that for v_r = 20 000 and $z = 9$, maximum velocity of a rocket in a free space will be ϑ_{max} = 2.3 (20,000) lg (10) ≅ 46 km/s.

To develop gaseous-phase reactors for the practical purposes of employing them in jet engines *shall mark an advent of a new era as far as interplanetary communication and mastering of the circumterrestrial outer space is concerned.*

Even with the already tested combination propellants ($F_2 + H_2 + Li$) maximum velocity of a single-stage rocket will be ($z - 9$) ϑ_{max} = 13,800 m/s, i.e. higher than the escape velocity (ϑ_r = 11,189 m/s).

This demonstrates that in years to come, single-stage rockets shall play a decisive role in mastering the circumterrestrial space (with due regard for losses due to gravity and resistance exerted by terrestrial atmosphere). We think that the major achievements or "milestones" in cosmonautics are as follows:

1. Launching of the first man-made satellite *Sputnik I* (1957, U.S.S.R.);
2. Flight made by Yuri Gagarin—the world's first flight ever made by a man at space velocity (1961 U.S.S.R.);
3. The first space-walk completed by a man (A. A. Leonov)—flight made by P. I. Beljaev and A. A. Leonov (1965, U.S.S.R.);
4. Launching of the first geostationary satellite of the Earth (1963, U.S.A.);
5. First men on the Moon (N. A. Armstrong and B. Aldrin) (1969, U.S.A.);
6. Launching of the first long-term orbiting stations (1971, *Salyut 1*, U.S.S.R.; 1973 *Skylab*, U.S.A.);
7. Successful flights of automatic interplanetary stations to the nearest planets of the solar system—Venus and Mars (1961-1977, U.S.S.R. and U.S.A.);
8. Creation of the world's first scientific and research space complex "*Salyut 6*", *Soyuz 26* and *Soyuz 27* (1977-1978, U.S.S.R.);
9. The longest flight (185 days) of the orbiting scientific and research complex *Salyut 6/Soyuz* (pilot-cosmonauts L. I. Popov, V. V. Ryumin).

Tsiolkovsky understood that his works were the source of revolutionary transformations in both science and engineering. In 1924 he wrote: "We witness that European science literally supports my conclusions both on a full possibility of space travel, and a possibility of setting up there [in outer space] dwellings, inhabiting the solar system. The latter yields two billion times the solar energy received by the Earth. Things are flaring up, and I have sparked off that fire."[17]

[17] Konstantin Eduardovich Tsiolkovsky, *The Rocket to Space* (Kaluga: 1924), p. v.

We should like to specifically emphasize the breadth and versatility of the talent possessed by Tsiolkovsky, as well as a great variety of scientific and engineering problems analyzed by him. An idea of the many-sided creative work done by this scientist and thinker can be gathered from the four volumes of the Collection of his works (published by the U.S.S.R. Academy of Sciences, 1951-1964).

Tsiolkovsky clearly understood the importance of his major works for the further advance of science and engineering. He saw "seductive and important prospects" for the advance of rocket engineering and cosmonautics. He knew that experimental aerodynamics is quite an indispensable component of the successful development of aviation and aerostation. He had discovered a reliable and strictly scientific way of mastering the cosmos and the circumterrestrial space. His research in the theory of multi-stage rockets has shown how to achieve space velocity of flight while employing commonly used propellant combustibles and oxidizers. During the last years of his life he published a variety of works relating to high-velocity aviation, and it was he who uttered the prophetic words that the era of propeller-driven aircraft shall be followed by the advent of the era of jet aircraft or "stratospheric" aircraft; he also anticipated the idea of developing vehicles riding on air cushions.

His astute mind was also interested in many problems of biology. In his remarkable paper "Plants of the Future," he pointed out that the major factor identifying the value of a plant used by a man for nourishment is the process of assimilating solar energy. Assuming that the corn yield provided by one hectare is one ton, then, as estimated by Tsiolkovsky, it turns out that the harvest would comprise but 1/6000 of energy received by the hectare from the Sun per year. In his writing "Plants of the future" he presented the following observation: "Wheat utilizes 0.016 percent energy received from the Sun. Banana utilizes 1.6 percent, i.e, 100 times that of wheat. The Burbank cactus utilizes up to 29 percent."[18]

In addition, Tsiolkovsky outlined a far-reaching program of mastering solar energy by means of *new* plants specifically developed for the purpose. He emphasized the poor efficiency or imperfectness of plants known at present, and placed a specific significance on the fact that they waste a lot of solar energy to overheat leaves, fruits, trunk, and branches.

He argued that the atmosphere of the Earth is poor in carbon dioxide (CO_2), which is necessary for growing plants and ensuring their good crop capacity; from the above follows the importance of designing green houses with a high CO_2 percentage.

---

[18] Konstantin Eduardovich Tsiolkovsky, "Plants of the Future," in his *Collection of Writings* (Moscow: U.S.S.R. Academy of Sciences, 1964), 4: 286-294.

We must point out unfortunately, that biologists do not pay enough heed to these writings of Tsiolkovsky. Academician A. E. Fersman, remarked, for instance:

> The article is of interest in its analysis given to the nature and life of plants, and in the emphasis made on those negative aspects which must be improved with respect to the energy equilibrium and chemical behavior of vegetable organisms . . . A brilliant idea which, as I presume, had not been ever set forth in this way prior to Tsiolkovsky.[19]

Tsiolkovsky was a meticulous and single-minded researcher. Indeed, his associates were amazed at his passionate animation and flights of creativity, his obsession with whichever problem was under study. He used to joke that deafness saved him from mean-spirited irritations generated by his surroundings, but allowing for overstatement, that may have helped him to immerse himself into his own reflections so thoroughly. He possessed a good memory, an encyclopedia-like mind, and a marvelous capacity for work. In studying his writings one is startled with both his strength of thinking and of feeling. His 1913 statement seems a fitting testament of his activities: "Life presented me much grief, and it was only my soul, which was boiling with a joyful world of ideas, that helped me overcome these misfortunes."[20] He worked daily, notwithstanding some severe life situations: poverty, loss of his near ones, mockery on the part of his colleagues/teachers and others in the streets of Borovsk and Kaluga towns, hunger and cold suffered during the Civil War, inability to publish writings ready for publication, through which writings he strived to bring his dearest ideas to his people.

Four months before his death, Tsiolkovsky broadcast to the Soviet Union a statement about his hopes for the future. He said: "Heroes and bold spirits will pioneer the first space route: Earth-orbit of Moon, Earth-orbit of Mars, and still further: Moscow-Moon, Kaluga-Mars." Since that time, many of the ideas presented by Tsiolkovsky have been taken up by scientists and engineers and have helped to chart the present course of the Soviet Space program.

---

[19] A. E. Fersman, "Analysis of General Writings by K. E. Tsiolkovsky," in *K.E. Tsiolkovsky, Collected Writings*, 4: 423-451, quote from pp. 335-346.

[20] Konstantin Eduardovich Tsiolkovsky to J. I. Perelman, 8 September 1913, in Archives of the U.S.S.R. Academy of Sciences, f. 796, reg. 3, file 23, pp. 1-2.

Chapter 15

Nikolai Alexeyevich Rynin (1877-1942), Soviet Astronautical Pioneer: An American Appreciation[1]

Frank H. Winter[2]

Nikolai Alexeyevich Rynin is little known outside his native U.S.S.R., save for his masterpiece, a nine volume encyclopedia of space flight and the world's first space encyclopedia, *Mezhplanetnye soobshcheniya* (Interplanetary Communication), published in Moscow and Leningrad between 1928-1932. Fortunately for western historians of spaceflight, the former historian of the U.S.'s National Aeronautics and Space Administration (NASA), Eugene M. Emme, arranged for this work to be translated into English at the suggestion of Derek de Solla Price, Professor of History of Science and Technology at Yale University. Rynin's books appeared as NASA TT's (Technical Translations) F-640 to F-648, and they were published from 1970-1971 in Jerusalem by the Israeli Program for Scientific Translations.[3]

Prior to this time, perhaps a handful of copies of this invaluable work existed in the west. During his lifetime Professor Rynin bestowed copies of the original Russian-language editions of the encyclopedia, as well as copies of his many other writings, to fellow space travel advocates in other countries, such as Willy Ley, the well-known German-born popular writer of the topic. In exchange, Professor Rynin would receive space flight literature from his western contemporaries. In a sense, this exchange was the first example of east-west space cooperation. It was by this means also that Rynin's

---

[1] Presented at the Fifteenth History Symposium of the International Academy of Astronautics, Rome, Italy, 1981.

[2] Department of Space History, National Air and Space Museum, Smithsonian Institution, Washington, D.C., U.S.A.

[3] Interview, Dr. Eugene M. Emme, 15 July 1981.

encyclopedia became more widely known outside his country, although the man himself and his considerable other accomplishments remained largely hidden. Rynin donated a set of books to the old Prussian State Library in Berlin, and the work was reviewed by German aeronautical journals like *Der Flug* and *Zeitschrift für Technik und Motorluftschiffahrt*. Willy Ley also mentioned them in his *Grundriss einer Geschichte der Rakete* (1932) and in subsequent writings.[4]

Figure 1 Professor Nikolai Alexeyevich Rynin (1877-1942). Portrait from Werner Brügel, *Männer der Rakete* (Leipsig, 1933) and copied as Smithsonian photo 76-16742.

[4] Willy Ley, "A Famous Russian Encyclopedia on Astronautics," *Journal of the British Interplanetary Society*, 13 (November 1954): 362-363; Willy Ley, *Grundriss einer Geschichte der Rakete* (Berlin: Hachmeister & Thal, 1932), p. 15; "Buchbesprechung," *Der Flug*, 10 Jahrg., 4 (April 1928): 140; *Zeitschrift für Flugtechnik und Motorluftschiffahrt*, 20 Jahrg., 7 (1929): 174.

Rynin himself wrote a rare autobiographical sketch which appears in Werner Brügel's *Männer der Rakete* (1933), although this book became scarce and was not well known outside Germany. English-speaking modern countries mainly knew of Rynin and his work through the English-language editions of Willy Ley's highly popular *Rockets, Missiles, and Space Travel* and other writings. The encyclopedia was also thoroughly reviewed (except for the first volume which was unavailable to the reviewer) by G. V. E. Thompson, in a three-part article in the *Journal of the British Interplanetary Society* for July and November 1954 and March-April 1956. Because Thompson made such an excellent survey, and since the English translations have been available for several years and were also reviewed, it is not necessary to treat the contents of the encyclopedia in any great detail here other than in a cursory outline. Rather, the present paper will concentrate upon the career of N. A. Rynin. It is interesting to note in this respect, that in his introduction Thompson wrote: "Little is known of the author, Professor Nikolai Alekseevich Rynin, except that he lived in Leningrad and was therefore possibly on the staff of Leningrad University. He was one of the leaders of the pre-war Russian rocket society, Gruppa Izucheniya Reaktivnogo Dvizheniya [GIRD]. Since he was publishing works on descriptive geometry and optics in 1907-1908, he must be at least 65-70 years old if still alive today." When Thompson wrote those words, in 1954, Rynin had been dead for 12 years. He died in Kazan in 1942 at age 65.[5]

Early Career

Nikolai Alexeyevich Rynin was born on 23 December 1877 in Moscow, the son of a minor official of the War Department. The Rynin family remained there until 1886 when Rynin's father died. The boy, with his sister and mother, first moved to Tiflis, then in 1887 to Simbursk (later Ulyanovsk). Here he spent eight years and received his secondary education at the Gymnasium. "My love of mathematics originated [here]," he later recalled, "and developed more and more, thanks to the influence of my teacher, E. Vilkovisky." Following graduation at the age of 19 in 1896, Rynin took and passed a competitive examination in order to enter the Institute of Transportation Engineering in St. Petersburg (later, Leningrad). The young man excelled in his studies and graduated in 1901. He was accepted as a graduate engineer with the Nikolayev Railroad and remained with the railroad until 1912.[6]

[5] N. A. Rynin, "Professor Nicolai Aleksejewitsch [sic] Rynin," in Werner Brügel, ed., *Männer der Rakete* (Leipzig: Hachmeister & Thal, 1933), pp. 79-85; G. V. E. Thompson, "A Famous Russian Encyclopaedia [sic] of Astronautics—1," *Journal of the British Interplanetary Society*, 13 (July 1954): 193-202; Part 2 in *Journal* (November 1954): 301-302; and Part 3 in *Journal* (March-April 1956): 82-91; T. F. Mackenzie, "Interplanetary Flight and Communication," *Spaceflight*, 15 (December 1973): 478; Colin A. Ronan, "Interplanetary Flight and Communication," *Journal of the British Astronomical Association*, 83 (August 1973): 390-392.

[6] B. Tsirkov, "Pilot, ucheniy, populyarizator," *Aviatsiya i Kosmonavtika* (1968), pp. 6, 22-23; Rynin, "Professor," 79.

In the meantime, Rynin was such a promising individual that he was sent abroad to advance his studies in engineering and to learn teaching techniques. He visited universities in Norway, Sweden, Denmark, Germany, and the U.S. Rynin's American travels were extensive enough and so fascinating for him that he afterwards wrote a 103-page illustrated paper entitled *Puteshestvive po Severo-Amerikanskim soyedinennym shtatam* (Travels Throughout the United States of North America). This paper was presented at the meeting of the Highway Engineers Society on 28 January 1905.[7]

It was neither Rynin's first publication nor his first lecture, because when he was with the Nikolayev Railroad, he was advancing his education and also lecturing, at St. Petersburg's Polytechnic Institute as well as the Institute of Transportation Engineering. At first he concentrated on the fields of civil engineering and descriptive geometry. In 1902 the Polytechnic Institute published his *Sbornik . . . orthogonal'nykh proyetsiyakh* (Collection of . . . Orthogonal Projections). He wrote about ice-breaking ships in 1903, (*Ledorezy*, or *Icebreakers*); metal-coating on constructions in 1905; a treatise on descriptive geometry in 1907; and a handbook on illumination in buildings in 1908. During his actual work career at this time, Rynin designed buildings, piers, and bridges. This work was carried out at St. Petersburg. Rynin was also preparing a dissertation entitled "The Calculation of Hinged Rings Made of Hard Materials," which was successfully defended in 1909 in the presence of the famous Russian mechanical engineer Ivan Vsevolodovich Meshcherskiy.[8]

It would have appeared as if Rynin's career as an engineer and teacher was set, but according to his later biographical sketch, "I first became interested in aviation in 1907." Rynin characteristically thrust all of his energy into the new field, voraciously reading and collecting everything he could. In the same year he went to France to see the Aeronautics Exposition at Nancy. From there he hastened to another aeronautical exposition in Frankfurt-am-Main, Germany. In 1909, he went to the very important international aviation meeting in Rheims, France. This was really the first great aviation meeting of history, held from 22 August to 29 August of that year, and in which 38 airplanes competed including Curtiss, Wright, Farman, Voisin, Bleriot, and Esnault-Pelterie biplanes. Many of the early aeronautical pioneers themselves were present, including Robert Esnault-Pelterie. It is not known whether Rynin met Esnault-Pelterie at Rheims, but years later they corresponded as Esnault-Pelterie, or REP as he was called, also became an eminent astronautical and rocketry, as well as an aeronautical pioneer. In fact, Rynin presented a set of his encyclopedias to REP who donated them to the British Interplanetary Society in 1949; those particular volumes were used by G. V. E. Thompson to make his 1954-1956 review of the encyclopedia.[9]

Rynin witnessed other aeronautical events in Europe in 1909 and also visited aeronautical institutions and attended conferences. From this period Rynin began to become

---

[7] Bibliographic list of works by N. A. Rynin were provided to the author by E. O. Mayo-Znak, Section Head, Information and Bibliographic Section, The Lenin Library, Moscow, 26 November 1976; Tsirkov, 22.

[8] See note above.

[9] Tsirkov, p. 22; Rynin, "Professor," p. 80; Charles Harvard Gibbs-Smith, *Aviation: An Historical Survey* (London: Her Majesty's Stationary Office, 1970), p. 145; Thompson, 13 (July 1954): 194.

personally acquainted with many of the world's leading aviation personalities, such as Alexandre Gustave Eiffel (of Eiffel Tower fame), who also used his considerable engineering talents towards aerodynamic studies; fellow Russian Igor Ivan Sikorsky, who afterwards pioneered in the development of the helicopter; Gaetano Crocco of Italy; and Dmitri Ryabushinsky of Russia. The latter two, like Esnault-Pelterie, also subsequently became rocketry pioneers. Several rocketry or astronautical pioneers, including Rynin, initially got their start in aviation.[10]

From 1909 Rynin, with Professor N. N. Mitinskiy, helped to organize the first All-Russian Aviation Club in St. Petersburg. He also began to teach courses in aeronautics in the Aviation Club, which was itself a milestone in the history of aeronautical education in Russia. In the same year, Rynin established a Civil Engineering Institute and was the first teacher. It was also through his tireless efforts that an aviation museum was begun here and that an aerodynamic laboratory was erected, including a wind tunnel with which the students could test the aerodynamic efficiency of their airplane, dirigible, and even train designs. The great Nikolai Yegorovich Zhukovsky, or Joukowski as he is known in the west, considered the founder of modern aerodynamics and hydrodynamics, commented very favorably on Rynin's laboratory in his book *Teoriya Aviatsii* (Theory of Aviation). Professor Zhukovsky noted the "lively activity" of the laboratory and that "around it there was formed a student aeronautics club whose members were zealous workers" who, with Rynin, presented papers at All-Russian aviation congresses. Engineer B. Tsirkov summed up Zhukovsky's remarks when he wrote that, in effect, Rynin was the first Russian to popularize and disseminate aviation in his country. This was a worthy precedent to Rynin's later career as a space flight popularizer.[11]

In the meantime, Rynin began his extensive writings on aviation. In 1910, for example, he authored *Noveyshiye uspekhi vozdukhoplavaniya* (Latest Progress in Aeronautics) and edited *Opredeleniye vysoty poletov i nekotoryye meterologicheskiye nablyudeniya vo vremya I-oy aviatsionnoy* . . . (Determination of the Altitude of Flights of Airplanes and Certain Meteorological Observations During the First Aviation Week in St. Petersburg, 25 May-2 June 1910). In 1910-1911 he published *Kurs vozdukhoplavaniya* (Course in Aeronautics), the first Russian textbook for students. In 1911, for the All-Russian Aviation Club, there appeared Aeroplan Farmana . . . (The Farman Airplane and the "Gnome" Engine).[12]

Nor was Rynin's original love of mathematics forgotten. *Nachertatel'naya geometriya* (Descriptive Geometry) was first published in 1911-1914 with a continuation and an expanded version in 1916. In 1913 Rynin wrote about wind pressure on buildings as a result of tests at the Institute of Highway Engineers, and in 1914 he produced a series of three related papers: the effect of wind pressure on bridge girders and other structures; wind resistance encountered by trains in motion; and the function of snow shields on vehicles. These papers, as noted by the American magazine *Aero and Hydro*

[10] Tsirkov, p. 23; Rynin, "Professor," p. 80.

[11] Tsirkov, p. 23.

[12] Bibliographic list.

for 4 July 1914, were presented at the Third Russian Aeronautical Congress in St. Petersburg.[13]

Rynin was a man of action too. He was an avid and accomplished aeronaut. He set a Russian balloon altitude record of 6,400 meters (20,997 feet) in a flight of the balloon *Vasiliy Kornya* during 21-22 September 1910 from St. Petersburg to the Ellola wood, Finland, just to the north of Lake Lagoda. The flight was made with F. I. Odintsov as a passenger. Rynin was proud of this feat and described it fully in Volume VI of his *Interplanetary Flight* encyclopedia; this volume covers "Superaviation and Superartillery." The flight is also treated in detail in *Vozdukhoplavaniya* for June 1911. In 1911 Rynin flew in the airships *Sokol*, *Lebed*, *Astra*, *Golub*, *Duks*, *Yastreb*, and *Tseppelin*. Later, he obtained a pilot's license. By 1914 he was experienced enough to be asked to participate in the summer tests that year of the world's first four-engine bomber, the famous *Il'ya Muromets*, designed by Igor Sikorsky. This is apparently how he met Sikorsky.[14]

Rynin was also one of the first Russians to present a paper at an international aeronautical meeting. In 1911 he came to Italy to read his report "On Testing Propellers," at the Fifth International Aeronautical Congress. The paper was so well received that the organizational committee invited him to read another paper which was subsequently presented in 1913 in Ghent, Belgium. Following this particular congress, Rynin journeyed to England, where he visited the Teddington Aerodynamic Laboratory. He then went to Paris, where he assisted Alexandre Gustave Eiffel in his aerodynamic experiments.[15]

In 1916 Rynin was appointed Professor of Descriptive Geometry at the St. Petersburg Institute of Transportation Engineering, and in 1920 he organized a Department of Air Communications there. In the following year he was made professor of that department.[16]

Early Astronautical Work to 1930

Rynin dates his interest in rockets to 1917, as he was then concerned with an ever more "rapid means of transportation." This inevitably "led me to the problems of interplanetary communication," he later wrote. In actuality there are different accounts of how Rynin began his interplanetary flight studies. In his autobiographical sketch, he says that during his infancy "I was already very fond of fireworks, and particularly of rockets . . . When a student at the Gymnasium, I prepared powder rockets by myself.

---

[13] *Aero and Hydro*, 8 (4 July 1914): 168.

[14] N. A. Rynin, *Interplanetary Flight and Communication* (Jerusalem: Israeli Program for Scientific Translations, 1971), Vol. II, 6 (NASA TT F-645), pp. 172-174; Tsirkov, 23; N. A. Rynin, "Vserossiskii prazdinkh vozdukhoplavaniya," *Vozdukhoplavaniya*, June 1911, 376-389.

[15] Tsirkov, p. 23.

[16] Rynin, "Professor," p. 80.

But, I became acquainted with the theory of this problem in 1903 only when K. E. Tsiolkovsky's work, *Investigations of Space by Means of Reactive Devices*, happened to come into my possession. My systematic occupation with the rocket problem and with the theory of spaceflight date, however, from 1917 only, when I had to straighten out my library, which consists of books on aviation and which now [1933] includes more than 6,000 titles. While examining some of those books, I found several papers that were concerned with the problems of rockets and of communication with the planets, such as the works of Professor I. [van Vsevolodovich] Meshchersky, [Dmitri] Ryabushinsky, N. Dshukovsky, *et al.* I began studying their work and to systematize the existing material. At the same time I succeeded in obtaining the classical works of Esnault-Pelterie, Goddard, Oberth, Hohmann, and other specialists of reactive propulsion from foreign countries. Later, news concerning practical work and experiments with rockets of [Max] Valier, [Fritz von] Opel, [Fritz] Stamer, Oberth *et al.* reached Russia. All that gave me the idea of revising and editing the collected material in a series of books, since books on interplanetary communication were practically nonexistent in Russia."[17]

In Engineer B. Tsirkov's account of the life of Rynin, there is another version of how Rynin began his astronautical investigations. Tsirkov says that Rynin read Tsiolkovsky's classic paper, cited above, at the end of 1915, and that it had made an enormous impression on him. Tsirkov also says that Rynin had already been acquainted with Tsiolkovsky earlier, when the two corresponded with each other from 1911. Tsirkov also adds that the two actually met each other at the Third All-Russian Aeronautical Congress, which was held in May, 1914 at the Civil Engineering Institute. However, Rynin did not mention this earlier correspondence nor the alleged meeting in Volume VII of his encyclopedia, which is titled "K. E. Tsiolkovsky Life, Writings, and Rockets." Moreover, he clearly stated in his 1933 autobiographical sketch, as quoted above, that he read the Tsiolkovsky book in 1917 and in the same year began his systematic study of the interplanetary problem.[18]

It took Professor Rynin 11 years of collecting and organizing his material from 1917 until the appearance of the first volume of his encyclopedia in 1928. At first, there was little he could publish in a preliminary way, simply because there was so little accessible data available. As early as 1918, however, he did write a review of the 1881 reaction-powered flying machine of Nikolai Ivanovich Kibaltchich. Kibaltchich, who had made the bomb which assassinated Tsar Alexander II, drafted his concepts of rocket propulsion while in prison. Kibaltchich's notes remained hidden in the secret police archives for almost 40 years. Rynin's commentary on the Kibaltchich project, along with an introduction by P. Shchegolev, appeared in the journal *Byloe* (The Past) for October-November 1918. Rynin also expressed his own ideas of interplanetary flight in his fantasy novel *V Vozdushnom okeane* (In the Aerial Ocean), published in Moscow in 1924. In Volume II ("Spacecraft in Science Fiction") and elsewhere in his encyclopedia, Rynin alluded to his fictional electromagnetic "radio ship" which was to go to the Moon by

[17] *Ibid.*, pp. 80-81.

[18] Rynin, *Interplanetary Flight*, Vol. II, 4 (NASA TT F-643), pp. 36-43, and Vol. I, 2 (NASA TT F-641), pp. 95-96.

electromagnetic waves transmitted from powerful stations set up on Earth; the stations were to transmit either positive or negative magnetic forces. Theoretically, this scheme would work but would require enormous amounts of energy. It is farfetched from our perspective, but it does show Rynin's visionary imagination at a time when the rocket was nothing more than a firework.[19]

As the literature of space travel increased, particularly as the works of Tsiolkovsky, Fridrikh Tsander, Oberth, Goddard, and Esnault-Pelterie became more well known in the Soviet Union, there emerged many space travel enthusiasts, who were bound to unite into clubs and societies. Professor Rynin soon became one of the best known and most ardent supporters of this new space travel movement and was himself responsible for establishing one of its earliest organizations. He appears to have included the subject of interplanetary flight among his regular lectures at the Department of Air Communications of the Institute of Transportation Engineering, particularly from the late 1920s. In 1927, the Leningrad branch of the All-Union Association invited him to present a talk on "interplanetary communications," and from this period he began his more extensive writings in the field of space travel, and all aspects of rocketry, which culminated in his encyclopedia. In January 1928, for example, his article "Reaktivnaya bomba" (The Rocket Bomb) appeared in the magazine *Voina i Teknika* (War and Technology). In February 1928 the journal *Vestnik Znaniya* (Herald of Knowledge) carried his article "O mezhplanetnykh soobscheniyakh" (On Interplanetary Travel), while a little later in the year the same journal printed his report on the rocket-propelled car of the German experimenter, Fritz von Opel. In the following year Rynin described space stations in his article "Kosmicheskaya observatoriya" (Space Observatory) in *Mir Prikluchenii* (World of Adventures). This article was based on the book by the Austrian pioneer Hermann Noordung, *Das Problem der Befahrung des Weltraums* (The Problem of Travelling into Outer Space), which had just been published. Rynin's article is indicative of how well-informed he was of the latest developments in the world's astronautical literature besides rocketry experiments. It is likely that he wrote to, or attempted to correspond with, Noordung.[20]

Rynin's organizational activities in astronautics, or "cosmonautics," as the new science came to be called, began in 1928. In the fall of that year he created a "Department of Interplanetary Travel" at the Leningrad Institute of Transportation Engineers, just as he had set up a Department of Air Communications eight years earlier there, when the institute was known as the St. Petersburg Institute of Engineering. He was named Chairman of the new Interplanetary Travel Department, which must have been the world's first educational body devoted to this subject. K. E. Veiglin and Yakov Issodorovich Perelman, both of whom shared with Rynin the distinction of being among the earliest writers of space travel concepts, were designated as Rynin's assistants. Perelman was, like Rynin, particularly well known as a popularizer of space travel, even in the West. In 1929, Rynin's Department undertook modest experiments with small pyrotechnical rockets with the intent of eventually constructing "stratospheric" vehicles, or rockets capable

---

[19] Tsirkov, p. 22; Willy Ley, *Rockets, Missiles, and Space Travel* (New York: Viking Press, 1959), p. 91.

[20] Tsirkov, pp. 23-24; Bibliographic list.

of ascending to 100 kilometers (62 miles). Liquid-propellant systems were also planned. Also in 1929, Rynin expressed hopes for an international research institute of interplanetary travel. This latter desire was doubtless an influence of Rynin's earlier participation in international aeronautical organizations. However, neither the stratospheric rockets nor the international space travel institute came about, mainly because of a money shortage, according to Rynin, and also because of "the complex problems involved" and "pressing academic duties."[21]

For much the same reasons, the first space travel organization in the U.S.S.R., and perhaps the world, also prematurely ended soon after its formation in 1924. This was the Obschestvo po Izutcheniya Mezhplanetnykh Soobschenii, or Society for the Study of Interplanetary Communication, also known as OIMS. This group was started on the initiative of the Soviet rocket pioneer Fridrikh Tsander and was based in Moscow. It is not known if Rynin was a corresponding member, though he was certainly aware of it, as this organization is described in Volumes I and IX of his encyclopedia. In any event, it lasted less than a year. Rynin's later involvement in the formation of another space travel group, the Leningradskaya Gruppa po Izutcheniyu Reaktivnogo Dvizhenia (Leningrad Group for the Study of Reactive Motion, or LenGIRD), was much more successful than either OIMS or Rynin's Department of Interplanetary Travel, but by that time anyway (1931), there was much more general support for the space travel dream and for the development of the rocket.[22]

Besides propagandizing, teaching, and encouraging the organization of space travel groups, Rynin also worked towards the space travel goal through some most interesting biological experiments carried out in the spring and summer of 1930 in the Scientific Research Bureau of the Institute of Civil Aviation in Leningrad. These experiments are well covered in a separate paper in Volume VIII of Rynin's encyclopedia, as well as in Bulletin No. 1 of the Institut Grazhdanskogo Vozdushnogo Flota (Institute of the Civil Aviation Fleet). This article, written with one of the principal participants of the experiments, the physician A. A. Likhachev, was also separately printed as an offprint in 1931 and entitled *Effekt uskoreniya na zhivye organizmy* (Effect of Acceleration on Living Organisms). Rynin built two centrifuges for the experiments, one with a radius of 1 meter (3.28 feet) which could spin up to 300 rpm, while the other had a radius of 0.32 meters (1 foot) which could revolve up to 2,800 rmp. Accelerations of 100 g and 2,800 g were thus obtained, respectively. Smaller animals and also raw chicken eggs were tested in the smaller centrifuge, while larger animals were placed in the larger machine. Different kinds of insects (cockroaches and flies), carp fish (in water and out of water), frogs, birds (siskins, pigeons, and crows), mice and rats, rabbits, and cats were subjected to systematic testing on the centrifuges and were afterwards examined by a team of physicians under the direction of Professor Likhachev. The doctors were M. Likhachev, V. M. Karasik, A. M. Vasil'ev and A. A. Sergeev. Among some of the conclusions reached were that the smaller animals withstood far greater accelerations than the larger

---

[21] Rynin, *Interplanetary Flight*, Vol. III, 9 (NASA TT F-648), pp. 3, 177; G. V. Petrovich, *Development of Rocket Engineering in the USSR* (no publisher, no date), p. 13.

[22] Rynin, *Interplanetary Flight*, Vol. I, 1 (NASA TT F-640), pp. 18-19, and Vol. III, 9 (NASA TT F-648), p. 3.

animals; the animals could endure the effects of acceleration if cushioned or placed in a certain position in the chambers; and the duration of acceleration was a key factor in the effects of the acceleration. As noted by the reviewer of Rynin's encyclopedia, G. V. E. Thompson, writing in 1956, the experiments "have relevance both now and in the future" of astronautics. He added, "They gave useful material for when the time came to construct much larger centrifugal machines for use in experiments with human beings. Such machines have, of course, since been constructed and used in many countries."[23] (Figure 2).

ИНСТИТУТ ГРАЖДАНСКОГО ВОЗДУШНОГО ФЛОТА

Н. А. РЫНИН и А. А. ЛИХАЧЕВ

ЭФФЕКТ УСКОРЕНИЯ
НА ЖИВЫЕ ОРГАНИЗМЫ

Отдельный оттиск из вып. 1-го Сборника Трудов Научно-Исследовательского Бюро

ЛЕНИНГРАД
1931

Figure 2 Title page, N. A. Rynin and A. A. Likhachev, *Effekt uskoreniya na zhivye organizmy* (Effects of Acceleration on Living Organisms) (Leningrad, 1931). Smithsonian photo A4314A.

[23] *Ibid.*, Vol. III, 9 (NASA TT F-648), p. 173, and Vol. III, 8 (NASA TT F-647), pp. 337-340; Thompson, 15 (March-April 1956): 87-88.

Figure 3 Cover, *Luchistaya energiya v fantazirakh romanistov i v proektakh uchenykh* (Radiant Energy in Science Fiction and in Scientific Projects) (Leningrad, 1931). Smithsonian photo A4316A.

The crowning achievement of Rynin's astronautical work during the late 1920s to early 1930s was unquestionably his encyclopedia. Published from 1928 to 1932 in Leningrad largely at his own expense, the work consists of nine volumes:

I. *Mechty, legendy i pervye fantazii* (Dreams, Legends, and Early Fantasies). (1928). 110 pages, 82 figures. 2,000 copies printed.

II. *Kosmicheskie Korabli* (Cosmic Ships). (1928). 160 pages, 164 figures. 15,000 copies printed.

III. *Luchistaya energiya v fantazirakh romanistov i v proektakh uchenykh* (Radiant Energy in Science Fiction and in Scientific Projects). (1931). 153 pages, 65 figures. 1,000 copies printed (Figure 3).

IV. *Rakety* (Rockets) (1929). 216 pages, 186 plates. 80 copies printed.
V. *Teoriya reacktivnogo dvizheniya* (Theory of Reactive Propulsion). (1929) 64 pages, 22 figures. Number of copies printed unknown.
VI. *Superaviatsiya i superartilleriya* (Superaviation and Superartillery). (1929). 218 pages, 200 figures. 1,200 copies printed.
VII. *K. E. Tsiolkovskii, ego zhizn', raboty i rakety* (K. E. Tsiolkovskii: Life, Writings, and Rockets). (1931). 112 pages, 34 figures. 1,000 copies printed.
VIII. *Teoriya Kosmicheskogo polete* (Theory of Cosmic Flight). (1932). 358 pages, 123 figures. 1,000 copies printed.
IX. *Astronavigatsiya, letopis' i bibliografiya* (Astronavigation, Chronicle and Bibliography). (1932). 358 pages, 123 figures. 1,000 copies printed.[24]

It is noted that Volumes IV, V, and VI were printed earlier than Volume III. This is because, as explained by Thompson, "Rynin's conception of the work changed as it progressed." The nine books were actually issued in the order of I, II, IV, VI, III, VII, III, IX and VIII. Some of the changes in publication plans were no doubt due to the great amount of progress made in astronautical theory by 1930 and also in the development of rocket technology. Rynin also mentioned that publication dates were uncertain because of "financial reasons." In the final volume of the work, Rynin acknowledged that he had not only received—and translated for the first time into Russian—the writings of Oberth, Walter Hohmann, Willy Ley, Esnault-Pelterie, and Robert Goddard, but he had also benefited from contributions by more than 300 readers "from remote parts of the U.S.S.R., supplying remarks and additions and often describing their own valuable and interesting researches in the field of rocket propulsion and the history of interplanetary travel." Rynin was particularly indebted to Tsiolkovsky, the patriarch of cosmonautics in the U.S.S.R. Tsiolkovsky's letter of 11 June 1926, in which he enclosed his autobiographical sketch, was proudly published in Volume VII, which specifically deals with Tsiolkovsky, and which was planned by Rynin to appear on the pioneer's 75th birthday. It was, however, published a year earlier.[25]

In his own autobiographical sketch written in 1933, Rynin specifically pointed out that "the main result of my theoretical work" (in the encyclopedia) "was the determination of rocket weight for various rates of climb and types of fuel. In addition, I determined the effect of the starting place of the rocket and of the direction of the landing flight. For a more precise determination of the trajectory of a space rocket, I constructed—with the aid of astronomical tables in orthogonal projection, in accordance with the rules of descriptive geometry—the trajectories of all planets of the solar system and of the principal comets and asteroids." Rynin's trajectory studies appear in Volume IX. Another important theoretical contribution to the literature is Rynin's own interpretation of the theory of reactive motion, which is mainly contained in Volume V, which also brings together the theories of other pioneers up to that time. This volume was considered significant enough to be reprinted as paper No. 101 in the *Sbornik Instituta*

[24] Thompson, p. 90; the NASA translations of Rynin's encyclopedia.

[25] See note above; Rynin, *Interplanetary Flight*, Vol. III, 9 (NASA TT F-648), pp. 3-4, and Vol. III, 7 (NASA TT F-646), p. 2.

Leningradskii Inzhenerov Putei Sooshcheniya (Proceedings of the Leningrad Institute of Transportation Engineers) in 1929. Single offprints were also available.[26]

Generally, all of the reviews of Rynin's masterpiece have been highly favorable, both in Rynin's day and in our own time. G. V. E. Thompson writing in 1956, for example, concluded that Rynin's work "certainly represents a great achievement, whether one is considering it as a statement of the position reached by astronautical studies in the 1930s or as a bulky literary work. As an encyclopedia of astronautics it is still quite unchallenged." Colin A. Ronan, writing in the *Journal of the British Astronomical Association* for August 1973 on the publication of NASA's English language translations of the work, commented:

> This is an important history that is itself of historical significance, and NASA is to be congratulated on the foresight to do this, and the wisdom to give libraries at least the opportunity to buy complete, and so take advantage of, a valuable example of NASA's responsible attitude to the history of the notable aspect of modern technology that is vital to astronomy today.

Robert W. E. Lademann, writing in *Luftfahrt* (Air Travel) for 22 February 1929, said of Volume II, then released: "It is regrettable that this work is not written in German!"[27]

Willy Ley, however, later recalled that not only did he get copies, but that Rynin "also mailed sets of his work to Professor Oberth 'and to Vienna,' the latter meaning either Franz von Hoefft or Baron von Pirquet, possibly both. I know that he had sent one to Esnault-Pelterie. I had a second but incomplete set, received from Yakov I. Perelman, which I passed on to a young and enthusiastic member of the VfR [the German Rocket Society which ceased in 1933], who intended to become the Society's librarian. I don't know what happened to it." Ley also mentioned that a set was sent to the Prussian State Library and to the old (pre-Second World War) Bibliothek für Wehrwissenschaften (Library of the German Army). The German space travel popularizer, Werner Brügel, who included Rynin's biographical sketch in his *Männer der Raketen* of 1933, also got copies, as did the German experimenter Rolf Engel. In France a set was sent to Alexandre Ananoff, the Russian-born pioneer living and working in that country since the 1920s. Ananoff in turn presented the books to the American astronautical writer, Frederick I. Ordway, III, who during the 1950s was studying at the Sorbonne in Paris. Years later, when NASA authorized the English translation of the encyclopedia, it used the incomplete set of Willy Ley—these had been deposited with other books of Ley's library to the University of Alabama. In order to fill in the gaps, Ordway, who was then a professor at the university, supplied the missing volumes, which came from his set originally given to him by Ananoff. Still another set of the encyclopedia was bestowed upon the British Interplanetary Society when it was formed in 1933, but this set was incomplete. The Royal Aeronautical Society in London also received a set. It appears from the pub-

[26] Rynin, "Professor," p. 82; Thompson, 13 (November 1954): 301; Rynin, *Interplanetary Flight* Vol. III, 9 (NASA TT F-648), p. 172.

[27] Thompson, 15 (March-April 1956): 90; Ronan, "Interplanetary Flight and Communication," 391; Robert W. E. Lademann, "Die Raumschiffahrt in der zeitgenössischen Belletristik," *Luftfahrt*, XXXIII Jahrg. (22 February 1929): 64.

lished correspondence from Rynin to Robert H. Goddard that the American pioneer also got the encyclopedias, especially as the American had furnished the Russian with biographical data for use in the books.[28] (Figure 4).

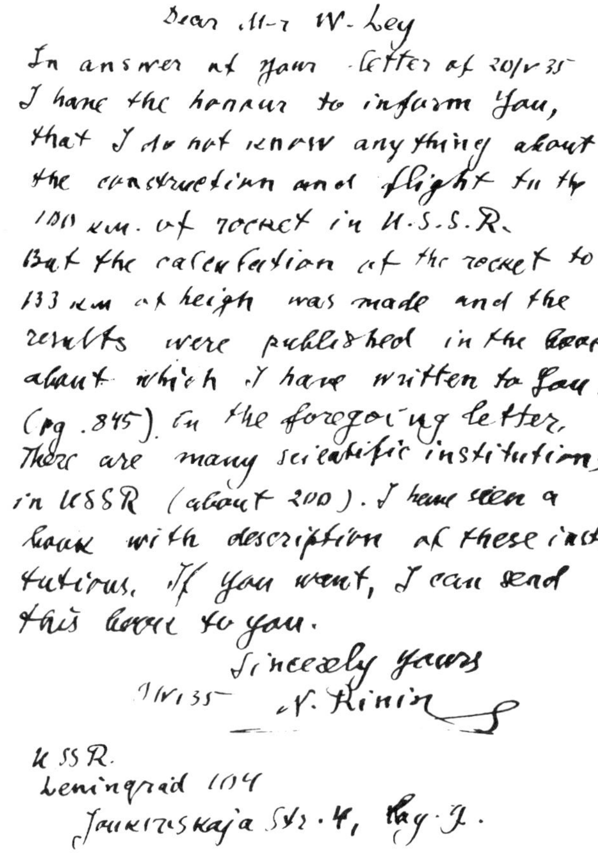
Dear Mr W. Ley
In answer of your letter of 20/V 35
I have the honour to inform You,
that I do not know anything about
the construction and flight to the
100 km. of rocket in U.S.S.R.
But the calculation of the rocket to
133 km of heigh was made and the
results were published in the book,
about which I have written to You.
(pg. 845). In the foregoing letter,
There are many scientific institutions
in USSR (about 200). I have seen a
book with description of these insti-
tutions. If you want, I can send
this book to you.
Sincerely yours
7/VI 35 N. Rynin
USSR.
Leningrad 104
Jamkirskaja Str. 4, kv. 9.

Figure 4 An example of Rynin's exchange of space and rocket information with the west, a letter to Willy Ley, 7 June 1935, in the Willy Ley Collection, National Air and Space Museum. Smithsonian photo 77-1260.

Later Work in Aeronautics

All the while Rynin that was pursuing his astronautical endeavors, he was still very much involved with aviation. Often the two fields overlapped. On 30 March 1927, for instance, a letter was written to Rynin from the Bureau for the Promotion of Inven-

[28] Ley, "Famous," pp. 362-363; Interview by Frederick I. Ordway, III, 21 July 1981; *Journal of the British Interplanetary Society*, 15 (September-October 1956): 92; Esther C. Goddard and G. Edward Pendray, eds., *The Papers of Robert H. Goddard* (New York: McGraw-Hill, 1970), I:575, 577, 583, 585, 607, 608.

tions of the Supreme Council of the National Economy of the U.S.S.R. requesting his judgment on the metal dirigible of the astronautical pioneer he admired so much, Konstantin E. Tsiolkovsky. He was also asked to report whether the airship designed by the great astronautical pioneer was suitable for official adoption by the Government. He responded with a positive report, in which Rynin also stated, "I believe that he [Tsiolkovsky] should long ago have been granted support in his further experiments." Rynin both revered and championed Tsiolkovsky and his ideas. The old pioneer clearly recognized this. "You . . . were the first to support my efforts," Tsiolkovsky told Rynin. "I, in fact, owe the most to you, to your writings, sacrifices, and boldness."[29]

In the early 1920s, Rynin performed invaluable aeronautical service in helping to lay out the Soviet Union's domestic air routes from the city of Rybinsk to Ul'yanov and from Rostov to Moscow. He also flew in a Fokker F.3 airplane to investigate the Moscow to Königsberg, Germany (now, Kaliningrad, U.S.S.R.), route, which also took him to Berlin and Potsdam. As a result of this work, the Soviet Union initiated its first international route. In 1931, Professor Rynin's career was further enhanced when he was named Chairman of the greatly expanded Department of Air Communications of the Leningrad Institute, which now became the Institute of Transportation Engineers. Rynin was also elected a member of the Institute's Academic Council.[30]

Throughout his career Rynin habitually kept abreast of the latest aeronautical developments. He also perennially wrote on aviation as well as space travel topics. These works ranged from his *Izmeritel'naya perspektiva i yeye primeniye v aerofotografii* (Measurement Perspective and its Application on Aerophotography), published in Petrograd in 1918, and *Kino-perspektiva i yeye primeniye v aviatsii* (Kino-Perspective and its Application in Aviation), printed in Leningrad in 1932, to *Zavoyevaniye stratosfery. K poletam sov. aerostratov v stratosferu* (Conquest to the Stratosphere. Towards Flights of Soviet Aerostats into the Stratosphere), and *V Stratosferu!* (Into the Stratosphere!), published in Leningrad and Moscow in 1933 and 1934, respectively. The latter books were essentially about the Russian stratospheric balloon flight of 30 September 1933 by G. Prokofiev, F. Godunov, and E. Birnbaum to an altitude of 60,695 feet (18,499 meters).[31]

As a former balloonist himself, the stratospheric leap of the balloon *USSR* and other flights excited Rynin very much. He was led to present the paper "Metody osvoeniya stratosfery" (Methods of Mastering of the Stratosphere), which was read at the First All-Union Conference on the Study of the Stratosphere, held in Leningrad from 31 March to 6 April 1934. The proceedings were afterwards published by the U.S.S.R. Academy of Sciences (Akademii Nauk SSSR) as *Trudy Vesesoyuznoi Konferentsii po Izucheniyu Stratosfery* (Proceedings of the All-Union Conference of the Study of the Stratosphere), published in Leningrad and Moscow in 1935. Rynin's paper was also published in Italian in the July 1934 issue of the aviation magazine *L'Aerotecnica.* It is interesting that at the same conference Professor A. A. Likhachev of the first Leningrad Medical Institute also presented a paper. This was on the biological experiments he had

[29] Tsirkov, pp. 25-26.

[30] *Ibid.*, 23.

[31] Bibliographic list.

performed with Rynin on Rynin's centrifuges four years earlier. Also present and listening intently was Sergei P. Korolev, future Chief Spaceship Designer of the U.S.S.R.[32]

It was perhaps because Rynin's paper appeared in *L'Aerotecnica* that the Italian rocketry pioneer Gaetano Crocco invited Rynin to give a paper at the important Fifth Volta Congress of High Speed Flight, held in Rome, from 30 September to 6 October 1935. But General Crocco had apparently already known of Rynin several years before, as Crocco's theories on what he termed "superaviation," or high-altitude, high-speed aircraft flight are well treated in Rynin's encyclopedia, particularly in Volume VI, which was largely devoted to aeronautics and is entitled *Superaviation and Superartillery* (1929). Professor Rynin, however, did not attend the Volta Congress. Instead, his paper, "Propulsione a reazione senza utilizzazione dell'aria esterna" (Reaction Propulsion without Utilizing Air), was read by Crocco. An abstract of the paper appears in *L'Aerotecnica* for September-October 1935, and was also published separately by the sponsor of the conference, the Reale Accademia d'Italia (The Royal Italian Academy), in 1935.[33]

Rynin's last major writing on aviation appears to have been his textbook *Proyektirovanive vozdushnykh soobshcheniy* (Design of Air Travel), published in 1937, although he also completed both a history and textbook on descriptive geometry in 1939. About 1941, not long before he died, Rynin also wrote a monograph entitled *Zavoyevaniye neba* (Conquering the Sky).[34]

Astronautical Work after 1930

Even before his space travel encyclopedia was complete, Professor Rynin had achieved considerable fame in his own country as an indefatigable and prolific worker in the cause of astronautics, although he tended to travel less during the 1930s period than previously. It was, therefore, a rare and honored occasion when he did pay a visit to Moscow to deliver a major lecture, illustrated with many slides, before the Air Force Academy of the Soviet Union on 28 November 1931 on "Reaction Propulsion."[35]

However, that very same month a much more important event took place. The famous LenGIRD (Leningradskaya Gruppa po Izutcheniyu Reaktivnogo Dvizhenia), or Leningrad Group for the Study of Reactive Motion, was formed. The idea for an organization in which dedicated enthusiasts could pool their talents to solve the great technical

---

32 F. J. Krieger, *Behind the Sputniks* (Washington, D.C.: Public Affairs Press, 1958), p. 353; N. A. Rinin [sic], "La conquista della stratosfera. 1° congresso dei Sovieti per lo studio della stratosfera," *L'Aerotecnica*, XIV (July 1934): 876-891; Yaroslav Golovanov, *Sergei Korolev: The Apprenticeship of a Space Pioneer* (Moscow: Mir Publishers, 1975), 282-284.

33 Alfred Bussemann to Frederick C. Durant III, 7 November 1974, copy in "N. A. Rynin" file, Library, National Air and Space Museum, Washington, D.C.; N. A. Rinin [sic]. "Propulsione a reazione senza utilizzazione dell'aria estera," *L'Aerotecnica*, XV (September-October 1935): 912-914; Ley, *Rockets*, p. 509.

34 Bibliographic list; Tsirkov, p. 26.

35 Golovanov, p. 244.

problems of perfecting a rocket for the conquest of the stratosphere, and then outer space, was not a new concept. Already, as mentioned earlier, the Society for the Study of Interplanetary Communication (OIMS) was created as early as 1924 but quickly became defunct before the year's end. MosGIRD (Moscow Group for the Study of Reactive Motion) was started officially on 18 August 1931 upon the initiation of Fridrikh A. Tsander and Sergei P. Korolev, and with similar groups, became much more viable. Through the aegis of the OSOVIAKHIM (Society for Assisting the Defense and Aviation and Chemical Construction in the U.S.S.R.) other branches soon sprang up at Leningrad, Kharkov, Kiev, Tiflis, and more remote places. "These groups," wrote Rynin in 1933, "have hundreds of members in every city." Rynin was one of the principal organizers of the Leningrad branch. The other organizers were the space travel popularizer Yakov Perelman and Vladimir Vasilyevich Razumov, then a newly graduated engineer who had worked in the Kronstadt shipyards, and another engineer by the name of M. V. Gazhala.[36]

Actually, LenGIRD was formed quite independently of MosGIRD but became a branch of the latter. Thirty years after these events, V. V. Tazumov recalled that at the end of March, 1931, "I was with Rynin, discussing with him the calculation of rocket flight by a method which I had proposed, and casually mentioned to him that I had entered with the OSOVIAKHIM and would be working in an unofficial way on rockets with the Bureau of Air Technology. Rynin responded with the fact that he had also spoken with the OSOVIAKHIM on this question and that he, Gazhala, and Perelman had joined together to participate in this work. Many times we gathered either at Rynin's or at my place to discuss at length, present propositions, and draw up plans regarding the development of work on rockets and the prospects of interplanetary communication. Thus was created the pioneer group for studying reactive propulsion at the Leningrad OSOVIAKHIM comprised of Rynin, myself, Gazhala, and Perelman."[37]

Razumov was named chair of the new organization, Perelman was vice-chair and Rynin was director of publicity. This was a natural extension of what he had already been accomplishing on his own. But it was no less an important task, for apart from building rockets, the group actively sought to educate the general public and also the scientific community on the potential of the rocket and the possibilities of spaceflight. Before the Second World War, LenGIRD built and flew several rockets, though it is not know if Rynin also played a role. In any case, according to Rynin, writing in 1933: "The propaganda for and the development of the idea of space flight have turned out to be the principal mission of GIRD." To reach this aim, for which Rynin was largely responsible, several broad programs were organized, including: popular public lectures which featured demonstrations, drawings, slides, and so on; seminars, consisting of some for a relatively uneducated audience such as workers and secondary school students, and other

[36] "Moscow Group for Study of Reactive Motion," and "Razumov, Vladimir Vasilyevich," in G. V. Petrovich, ed., *The Soviet Encyclopedia of Space Flight* (Moscow: Mir Publishers, 1969), pp. 270 and 333, respectively; Rynin, "Professor," p. 85.

[37] V. V. Razumov, "From the History of the Leningrad Group for the Study of Jet Propulsion (LENGSJP)," [sic], in USAF Systems Command, Foreign Technology Division, *From the History of Aviation and Cosmonautics (Collection of Articles)*, FTD-HT-23-1233-68, p. 20.

seminars for more technically trained or professional people. Systematic courses were also arranged for students at technical universities; specialized treatises were published, including several of Rynin's writings; and newspaper and magazine articles were written to further draw public interest.[38]

Prior to Rynin's association with LenGIRD, he was already associated with space travel societies in the West. Rynin, the internationalist who early advocated an international space travel organization, donated money to the German Rocket Society, according to donor lists in the Society's journal *Die Rakete* (The Rocket), as early as 1929. Also, by January 1931 he was corresponding with Noel Deisch of the American Interplanetary Society (later called the American Rocket Society). It was through Deisch, who had won an honorary mention in the first REP-Hirsch Prize of Astronautics in 1928 for his paper "The Navigation of Space," that Rynin was introduced to *The Bulletin of the American Interplanetary Society*. Deisch, who was then living in Washington, D.C., arranged for Rynin to receive a subscription to the magazine which was published in New York. Deisch also led Rynin to become an honorary member of the American Interplanetary Society. Later, Rynin was similarly named an honorary member of the British Interplanetary Society after it was formed in 1933.[39]

In the Soviet Union, Rynin's name became more intimately linked with that of his idol Konstantin Tsiolkovsky when, on 17 October 1932, Rynin was one of the main speakers on the celebration of the 75th anniversary of Tsiolkovsky's birth. Rynin's words must have touched many. Earlier, he had told Tsiolkovsky, ". . . Your generous heart will be valued by posterity . . . Your beautiful works and the nobility of your feelings will make your name immortal." Perhaps the same could have been said of Rynin. Through his 255 articles and books on several different topics throughout his lifetime, his many years of devoted teaching, and his passionate lectures on spaceflight, Rynin must have influenced literally thousands. This was a generosity too, as he gave of himself to a noble cause, the enlightenment of mankind.[40]

In July 1942 Nikolai Alexeyevich Rynin, along with many other Leningraders, was evacuated from the city under siege by the invading Germans. He was now seriously ill. Soviet military pilots transported him to Yoshkar-Ola and from there to the hospital in the ancient city of Kazan, more than 700 miles (1,126 km) south of his beloved Leningrad. Where many years before he would have delighted in such an aerial adventure, now this plane trip was apparently too arduous for his failing health. Rynin died of lung cancer at Kazan on 28 July 1942 and was buried with military honors in the Arsk Cemetery.[41]

Rynin was hardly forgotten. Not long after the war, according to a lengthy tributary article in *Pravda* on the 100th anniversary of the birth of Rynin, in December 1977,

---

[38] N. A. Rynin, "Die GIRD," in Werner Brügel, *Männer der Rakete*, pp. 138-39; "Leningrad Group for Study of Reactive Motion," in G. V. Petrovich, *The Soviet Encyclopedia*, pp. 217-218.

[39] Frank H. Winter, "Birth of the VfR: The start of Modern Astronautics," *Spaceflight*, 19 (August 1977): 247; "John Noel Deisch" File, Library, National Air and Space Museum, Washington, D.C.

[40] Tsirkov, pp. 22-26.

[41] *Ibid.*, p. 26.

a young man knocked at the front door of No. 4 Zhukovskaya Street in Leningrad where the professor had lived since 1932. A woman answered but informed the young man that the professor had passed away. The young man who had been so much inspired by the professor was identified as Georgi Grechko, the future cosmonaut of *Soyuz 6* and *Soyuz 26*, and who lived in space aboard the space stations *Salyut 4* and *Salyut 6*. Still another young man influenced by Rynin and who in fact studied under him was B. V. Raushenbakh, an early GIRD member, co-worker of Sergei P. Korolev, and later an academician and supervisor of the control systems for the Soviet Union's first spacecraft to reach the Moon, the Luna series.[42]

Surely the greatest remembrance and monument to the works of Rynin was the immortality bestowed upon him by the International Astronomical Union, which named a crater on the far side of the Moon in his honor.[43]

---

[42] "Avtor 'Mezhplanetnye soobsheniya' k 100-letino so dnya rozhdennya N. Rynina," *Pravda*, 2 December 1977; Interview by Dr. Eugene M. Emme.

[43] "Rynin, Nikolai Alexeyevich," in G. V. Petrov, *The Soviet Encyclopedia*, p. 36.

Chapter 16

The Contribution of Robert Esnault-Pelterie to Astronautics[1]

P. Contensou[2]

The pioneering work of Robert Esnault-Pelterie is distinguished by its exceptional scope in the field that he covered—the whole of aeronautics and of astronautics—the range of technology that he has enhanced, and the variety of his abilities as an engineer who was as gifted for theoretical work as for experiments.

To speak very rapidly of a subject which does not form part of the agenda of our Congress, we should simply recall that Esnault-Pelterie numbered among the brilliant young men of several nations who achieved the first powered flights at the start of this century. For this reason, the Aeroclub of France awarded him retroactively French pilot's license No. 4 when licenses were instituted in January 1909. He is known particularly to many of my compatriots as the inventor of the joy-stick, the lever with two degrees of freedom that allows a pilot to control simultaneously, with one hand only, the ailerons and the elevator of his aircraft. But a much more important side of his work related to the general design of the aircraft and of its engine, for example, the radial engine and the use of metal that he pioneered to build the airframe and the propeller.

It is remarkable to see that a man who had contributed so decisively to the strides made in aviation, and who might have devoted a glorious career to its development and progress, immediately looked further ahead. The bounds that man was starting to make successfully into the lower layers of the atmosphere were, in his eyes, only a stage in the great upsurge that would tear him from his natal planet one day and bear him up to the

---

[1] Presented at the Sixteenth History Symposium of the International Academy of Astronautics, Paris, France, 1982.

[2] President of the Association Aéronautique et l'Astronautique de France (AAAF), and member of Office National d'Etudes it de Recherches Aerospatiales (ONERA), Chatillon, France.

stars. This was the problem to which he was to devote his whole life in an environment of skepticism which was completely general, at least at the start. He presented a paper to the French Physics Association in 1912, which he had to camouflage under the low-profile title of (to use his own expression) "Considerations and Results of an Infinite Lightening of Engines," and in which he affirmed his confidence in the possibility of interplanetary traveling. He allowed his ideas to ripen for a long time and compared them with those of foreign rivals that he appreciated—Oberth, Hohmann, Goddard—and published in 1930 a major work "Astronautics," which is a veritable epitome of knowledge at the time in a field that greatly benefited by his original contribution. The variety of subjects that he dealt with, the accuracy of his views, and the reliability of his forecasts, today remain a subject of astonishment.

But Esnault-Pelterie was not only a brilliant theoretician. He was also an engineer passionately attached to proving the accuracy of his concepts by experiments. It was, in this case, the technique of the rocket which was the key to astronautics. It was the development of this technology that he was to devote his work during the greater part of the ten years between 1930 and 1940. He set up a test installation at Boulogne; government assistance, which was at first very limited, gradually developed thanks to the influence of one of his most faithful supporters, General Ferrié, who knew and appreciated him since the time of his military service, and whose prestige, as the creator of wireless telegraphy, opened many doors to him.

Esnault-Pelterie first used what we call today a single propellant, tetranitromethane. An accident that deprived him of three fingers of his left hand led him to prefer another propellant which he first believed to be less reliable, the liquid oxygen/petroleum ether pair. It was ambitious to approach cryogenic propulsion at the start, but Esnault-Pelterie overcame all the difficulties and rapidly succeeded in developing a rocket that delivered about 100 kg of thrust for one minute with a very reasonable specific impulse on the order of 230 sec.

The thermal problems were solved by cooling the engine with running water, and all the subsequent efforts of Esnault-Pelterie were directed to a more operational system, based either on liquid-oxygen cooling or on a non-cooled system that made use of highly refractory materials. He made the greatest efforts in this second direction but also encountered a number of difficulties—melting, perforation, galloping corrosion at temperatures where, as he said, "everything combines with everything else." Resolved not to be overcome by any difficulty, he invented electrical furnaces in order to make his own materials. When his personal activity ended with the Nazi invasion in 1940, he left a rich harvest of experiments to the men who were to continue his work, Professor Montagne and Colonel Barré. He himself followed the development of astronautics after the war, but as a spectator. Esnault-Pelterie saw the brilliant confirmation of his ideas in the launching of the first Sputnik, shortly before his death.

The Ideas of Esnault-Pelterie on Piloting and Navigation

Problems of piloting and navigation figure among the subjects on which the most personal and possibly the most original viewpoints are furnished by Esnault-Pelterie.

When examining the possibility of sending a projectile or an inhabited vehicle to the Moon, Esnault-Pelterie realized after a brief calculation of errors that the necessary precision would not be attained by the very strict heading and velocity imposed at the start. He understood that a corrective engine was needed, that it would operate on data supplied by a navigation system—and which might be the engine used later for moon landings. The engine had to be orientable but it would be simpler to use it to orient the vehicle itself. The nozzle was therefore mounted on a universal joint and operated by a control with two degrees of freedom—evidently a joy-stick—allowing the thrust to be offset in relation to the center of gravity in order to create the torque required for correct steering. He believed that an electric control system would be the best means of maintaining orientation.

Esnault-Pelterie then understood that it would be necessary to control attitude, even when the engine was not operating, either to make an astronomic observation or for reasons of habitability. This could be done by means of "small adventive rockets" that could be operated when required but, to economize fuel, it would be better, he thought, to use the effect of reaction of small electric motors fitted at right angles. Then compared with systems actually used today, this description only lacks the idea of the engine being essential to assist a motor that has reached its maximum speed when it has to counteract a disturbing torque of significant average value for too long a time.

Trajectory correction assumes that the exact position of the vehicle is known. Bearings taken on planets offered a possibility, but were of insufficient accuracy. Esnault-Pelterie then thought of the principle that we today call inertial navigation. He described the problem in particularly clear terms:

> Given a point "*m*" that is mobile from a position that we take as the origin and in relation to three axes of coordinates that we consider as fixed, is it possible to design an instrument which can indicate at any instant the position of the mobile point in relation to the axes, with no link, either material or immaterial, between the mobile point and any part of the system of fixed coordinates?

He immediately produced a theoretical solution to the problem:

> If we possess a means of linking, to the mobile point "*m*," a system of auxiliary axes of coordinate axes which remain parallel to the fixed axes while following the point in its movement, and three components of acceleration along the three axes, we shall know at each instant the values of:
>
> $$\frac{d^2x}{dt^2}, \frac{d^2y}{dt^2}, \frac{d^2z}{dt^2}$$
>
> and, if we possess a means of twice integrating these time functions, we shall know at any instant the coordinates of the mobile point in relation to the fixed system.

Esnault-Pelterie then reflected on the possibilities of designing the method practically. He immediately imagined several principles of an accelerometer that would carry out the first integration, that is, supply the components of the velocity vector.

—The first operates by viscosity. A mass caused to move along one of the axes is subject to a perfectly-linear braking force that is sufficiently powerful for its movement in relation to the mobile piece to remain very small when compared with the distance covered. Its final movement is proportional to the velocity.

—The second system is an off-balance gyroscope. The torque applied to the gyroscope is proportional to the acceleration and thus to the precession rate also. The angles through which it has precessioned are therefore proportional to the velocity.

—The third system is a little more complicated and is inspired by the classical measurement of gravity by the pendulum method, with the counting of oscillations supplying the desired effect of integration. Because the oscillation frequency is not proportional to gravity but to its square root, he had the idea of linearizing the system by making the apparent gravity act on a pendulum retained by a spiral spring and therefore with a natural frequency that the acceleration would only modify by a small quantity.

This natural frequency is finally eliminated by the use of two opposed pendulums, one with its equilibrium point stable and the other with its equilibrium point unstable in the direction of the acceleration component that it is desired to integrate. Each controls rotation of a gear train through an escape mechanism taken from clock systems. The outputs of the two systems act in reverse directions on a differential of which the spindle indicates the velocity attained by the vehicle, by the angle through which it has rotated since the start.

Esnault-Pelterie seemed to favor the latter system, because he immediately saw the advantage that he could gain from it to carry out the second integration. Let us replace, in an instrument based on the same principle, the force due to the apparent gravity by a force applied proportionately to a previously-measured velocity component, for example, by the action of a spiral spring of which the other end is moved in relation to the result of the first integration. The output of the second instrument will supply a position coordinate. A variant using a torsion pendulum, of which the stiffness is modulated proportionately to the variable to be integrated, appeared to him to exhibit certain advantages.

Esnault-Pelterie then gave attention to another aspect of the problem—the construction of a constant-orientation platform that would have to carry these various instruments. He recognized that it was possible to do so by means of two gyroscopes controlling electric motors, but this method appeared to him to be "enormously complicated." He then realized that acceleration of the vehicle is directed along the axis of the nozzle, and that it simply had to be measured along this direction. The orientation of a nozzle axis is itself reconstituted by double integration based on the information given by two angular accelerometers. (It is curious that he appeared to overlook the possibility of directly gauging the angular velocity by means of gyrometers and of reconstituting attitude by one integration only.) The method remained mysterious to him, and he admitted that he had not had the time to examine it more deeply.

He finally examined the problem posed by the existence of a gravitation field to which the accelerometer is insensitive, but which has a fundamental influence on gauging the trajectory. He very soon saw that, if the field is known as a function of position, it was simply necessary to add, at each state of integration, the acceleration as measured by the accelometer to the gravity acceleration calculated for the position immediately preceding and suitably extrapolated. He gave no indication of a mechanical device that would be able to cope with this complication. The problem can only be resolved, in effect, by means of the enormous facilities that were later supplied by on-board digital computers.

The ideas presented by Esnault-Pelterie on the subject of navigation are incontestably original in that he certainly did not take them from any of his predecessors. He would not have failed to have said so if he had, for he took great care to cite his sources. This of course does not mean that other workers in other countries did not have similar concepts at the same time, or earlier, and without his knowing. In any case, the history of astronautics and the technology of military missiles soon manifested their full value. The three principles of measuring acceleration that he proposed have been used effectively. The last even constitutes, in a slightly different form, one of the most advantageous present-day designs for an integrating accelerometer. He simply followed the evolution of the clock-maker's art. The hinged pendulum and spring is replaced by an elastic vibrating rod and the oscillations are counted by electronic methods.

The two methods of application of inertial navigation envisioned by Esnault-Pelterie (a stabilized platform or use of accelerometers directly connected to the structure) have both been used and are still today in competition. What a satisfaction would it have been for our pioneer if he had been able to anticipate that the two systems that he had imagined would be used in association, one as a normal system and the other as a standby system in the Apollo program, that was to realize his dreams, in flight to the Moon.

Conclusion

To recall the contribution of Esnault-Pelterie in a very specific sector, in order to illustrate the variety of his talents, I would not like to fail to speak of the fundamental role that he played in the promotion of international cooperation in the field of astronautics. We know that in 1929 he founded an international prize for astronautics with his friend, the banker Hirsch, and that his first gesture was to award it to Hermann Oberth. Yet, Oberth was a rival in the scientific field and, at the same time, a national of a country whose relations with France were still scarred by a recent and cruel war. Oberth expressed his gratitude in terms that remained printed in all the editions of the work that he published a little later, in spite of events that were again so tragic.

Esnault-Pelterie believed that astronautics is a subject sufficiently ambitious and fascinating to justify a coordinated effort by the whole of humanity, with the hope of perhaps turning man away from his traditional conflicts. This is the spirit that today animates our International Federation of Astronautics, the IAA, which has the duty of particularly honoring the memory of one of the most unquestionable originators of this noble ambition.

Chapter 17

Romanian Professor Elie Carafoli: Fifty-Five Years of Devotion to Modern Aeronautics and Astronautics[1]

Florin Zăgănescu[2]

Romanian Professor Elie Carafoli's professional activity produced more than 150 scientific works on fluid mechanics, aerodynamics and applied astronautics, many of them translated into the English, French, Chinese, German and Russian languages. His works, "High Speed Aerodynamics" (1956), "Wing Theory in Supersonic Flow" (1969) and "Fluid Mechanics" (1981-1982), were appreciated as reference works in this field of engineering. His contribution to aerodynamics and fluid mechanics embrace the profile theory, the finite-span wing theory, supersonic flow around various forms of wing and wing-fuselage assemblies; conic and quasi-conic motion theory etc. In his career he has received numerous orders and medals. He is a member of many academics and important societies, such as Honorary Fellow of the Royal Aeronautical Society-London; member of the International Academy of Astronautics of the International Astronautical-Federation etc. From 1961 Professor Carafoli led the Commission on Astronautics of the Romanian Academy and from 1958 the Technical Science Department of the Academy; in the period 1968-1970 he was president in action of the IAF. An aerodynamic airfoil class for aircraft wings and two performance aerodynamic wind tunnels are associated with his name, and Professor Carafoli is the founder of the Romanian school of aerodynamics; he is also a councilor at the Ministry of Machine Building Industry.

---

[1] Presented at the Sixteenth History Symposium of the International Academy of Astronautics, Paris, France, 1982.

[2] University Professor of Aerodynamics; and Secretary, Committee on Astronautics of the Academy of the Socialist Republic of Romania.

The Academy of the Socialist Republic of Romania recently celebrated the 80th birthday of the Romanian scientist of international repute, Elie Carafoli. A founder of the Romanian school of aircraft-building, Professor Carafoli, together with Traian Vuia, Aurel Vlaicu and Henri Coandă, comprise the famous quartet of Romanian aviation pioneers that ranks Romania among the countries that have made remarkable achievements in this field of science and technology in the twentieth century.

As a former student, and collaborator for more than thirty years, of academician Professor Carafoli, I am honored to say that this eminent professor's activity materialized in an impressive number of basic works in the mechanics of fluids, aerodynamics, applied aeronautics and astronautics and, also, in some remarkable achievements on aircraft manufacturing and in aerodynamic test units. I will consider the main aspects of Professor Carafoli's life and work, his activity being carried out with youthful energy for the delight and pride of his students and collaborators.

Working for Science and His People

Born on the 15th of September 1901 at Varia, near Salonic, Elie Carafoli was a member of a Romanian family which performed important tasks in the cultural life of their regional community. After graduating from the Romanian elementary school in his native town, he attended to advanced studies at the Romanian High School in Salonic until the beginning of the First World War, which prompted the young Carafoli to go back to Romania in 1916, bypassing areas upset by the conflagration in this region. After his arrival in Bucharest, he graduated High School and, with brilliancy, the Polytechnical Institute as an electromechanics diploma engineer. To complete his scientific education, Elie Carafoli attended, and graduated from, the Faculty of Sciences at the Sorbonne in Paris, obtaining his licentiate's degree in Physics and Mathematics in 1926, and, two years later, a doctorate in Physics, with a thesis entitled: "Contributions à la théorie de la sustentation en aérodynamique" (Contributions to the theory of sustentation in Aerodynamics). In this period the science of aerodynamics became the passion of his life, Carafoli being the founder of the Romanian aerodynamic school and teacher of an important number of specialists, whose subsequent activity was to place Romania in the forefront of research and practical achievements in aeronautics. This activity started in Paris 55 years ago, when Carafoli collaborated with the French professor Albert Toussaint, who was chairman of the Aerotechnic Institute at Saint-Cyr under the auspices of the Sorbonne. At the same time the young Romanian doctor served as an assistant-professor at the Fluid Mechanics cathedra, led by reputed Professor Paul Painlevé. The scientific papers and works published there (a lot of them with Professor Toussaint) brought him the repute of a specialist in aerodynamics.

In 1928, fresh from the Sorbonne, doctor Elie Carafoli returned to Romania and started a rich and fruitful career: he inaugurated the first aerodynamics and flight mechanics course at the Polytechnical Institute in Bucharest, Romania, which was thus among the first countries training aviation engineers; he designed and built (together with Professor Ion Stroiescu) a modern aerodynamic wind-tunnel, the first in the south-east of Europe (in 1930-1931). In this wind-tunnel were tested all the prototypes of the

Romanian designed and manufactured aircraft of the period 1931-1945 and, also, systematic tests were made on the sections of wings drawn to an original method proposed by Professor Carafoli, some of them being mentioned as "Carafoli sections" by the great scientist Milne-Thomson. At the same time, Professor Carafoli performed many technical actions in aircraft engineering, i.e., after 1928 a team of aircraft engineers (the Frenchman Lucien Virmoux included), designed and built, under Carafoli's leadership at the Romanian Aircraft Industry (IAR) factory at Brasov, many low-wing monoplanes (as IAR CV-11, IAR-14, IAR-15, IAR-16, etc.), a daring formula at the time. As one of the directors of this factory, Carafoli built new and modern research and project departments for it. Also, the founding near Ploiesti City of an important factory manufacturing accessories for aviation, named IRMC (Romanian Mechanical and Chemical Industry), is associated with his name.

In 1948, Professor Carafoli was elected a full member of the Romanian Academy and received the task of organizing the scientific research in applied mechanics; for two decades, successively, as chairman of the Institute on Applied Mechanics, and as chairman of the Institute of Fluid Mechanics, and last as honorary chairman of the Institute of Fluid Mechanics and Aerospace Buildings, professor Carafoli trained many researchers, the results being determinants for the practical use of mechanics in the national economy. As proof, there are many scientific contracts carried out and implemented in various branches of the Romanian economy by researchers of the institutes headed by Professor Carafoli. The scientific reports and papers of these researchers were published in journals such as, *Studies and Researches in Applied Mechanics* (in Romanian) and *Applied Mechanics Review* (in other languages), Professor Carafoli being their permanent editor-in-chief.

Academician Elie Carafoli made major contributions to the popularization and acknowledgement of the great merits of Romanian forerunners in aviation and astronautics. In 1961 the International Astronautics Federation welcomed the Commission on Astronautics of the Academy of the Socialist Republic of Romania as a full member, while Professor Carafoli was elected to positions of high responsibility, including the chairmanship of this international body (1968-1980).

High Scientific Level Works

Professor Carafoli's large and diverse scientific work reflects his constant and efficient preoccupation with directing the basic science to its practical applications. We must chronologically specify the results of the aerodynamic sections theory, the method to draw and, also, the important class of sections named "Carafoli Sections" (with round trailing edges). In his studies referring to formations around a contour, Professor Carafoli was the first to solve the motion problem of a general shape section rotating around a fixed point. Together with the research of the winglets influence on the aerodynamics of a wing, Professor Carafoli was the first to establish the formula of the forces and couples of hinges acting on the winglet and/or on the moving part of the tailplane. The tests carried out at the Aerotechnic Institute at Saint-Cyr (France), and later at the polytechnical Institute wind-tunnel at Bucharest, confirmed the section theory and the finite-span wing theory. Professor Carafoli drew up a general method to study these wings and the general solution of the variable incidence wing problem based on it, and he established the lateral command winglets theory and the turning wing theory.

Another ample work entitled "Aerodynamics"was translated into German and Russian, including important and efficient solutions, i.e., for the minimum drag at the generalized lift systems, of the wings with central fuselage, etc. After 1952, Professor Carafoli oriented his research to high-speed aerodynamics, where he carried out, and then published, many ideas in the basic handbook *High Speed Aerodynamics* (1956), translated into English and Russian. In 1969 at Pergamon Press, Professor Carafoli edited, with some of his collaborators, the prestigious book *Wing Theory in Supersonic Flow*, including some important results obtained in a supersonic wing study based on the conic motions of a higher order; the first theoretic solution for cross wing problems; the straight wing with normal lateral discs; the complete study of the supersonic wing with conic fuselage, etc.

Professor Carafoli had the merit to attract at such modern studies, in the laboratory or at the working desk, a number of collaborators, giving them the possibility to become well-known abroad. Many actual problems of supersonic aviation were solved by the Romanian aerodynamic school, Professor Carafoli's collaborators using his theoretical model of the supersonic conic motion by separating the subsonic flow at the leading edges of the wing. For supersonic motion and, starting from it for moderate-hypersonic, Professor Carafoli established a unitary formula for the pressure coefficient, to be used in a large range of motions. Together with his collaborators, Professor Carafoli studied, in addition, the round cone having or not havinf an incidence in supersonic or moderate-hypersonic motions; the effect of the lateral plane jets ejected from the ends of a small aspect ratio wing; the combustion waves in supersonic velocity; the variable geometry wing, etc.

As chairman, moderator or lecturer at many international congresses and symposia, Professor Carafoli contributed to Romania's prestige abroad and to international scientific cooperation. His continuous preoccupation with aerodynamic and fluid mechanics is today also confirmed by his rich editon program, i.e., a new, ample handbook/monograph entitled *Dynamics of Incompressible Fluids* (1981-1982). Professor Carafoli's works were mentioned as reference works in many foreign handbooks on fluid mechanics and aerodynamics, his original scientific ideas and solutions receiving praise from great personalities in aeronautics. Some of his solutions, moreover, served as starting points for the master of science thesis preparation in our country and abroad.

We can make a brief overview of Professor Carafoli's important handbooks and main groups of published works.

General handbooks, and monographs:

1. *Aérodynamique des ailes d'avion* (Aerodynamics of the aircraft wings), Librairie Chiron Editeur, 40 rue de Seine, Paris, 1928.
2. *Théorie et tracés des profils d'ailes custentatrices* (Theory and designs of the portant wing sections), collabor, with A. Toussaint, Librairie Chiron Editeur, 40 rue de la Seine, Paris, 1928.

3. *Influence des ailerons sur les propriétés aérodynamiques des surfaces sustentatrices* (The winglets influence on the lift surfaces aerodynamics), Publications de l'Aéro-Club de France, Centre de Documentation Aéronautique Internationale, 1931.
4. *Recherches expérimentales sur ailes monoplanes* (Experimental researches on monoplane wings), Publications scientifiques du Ministère de l'Air, Librairie Gauthier-Villars, Paris, 1932.
5. *Travaux du Laboratoire Aérodyamique* (Aerodymamic Laboratory Review), Publications scientifiques de l'Ecole Polytechnique de Bucharest, Imprimeriile Statului, Bucharest, 1938.
6. *Théorie des ailes monoplanes d'envergure finie* (Finite-span monoplane wings theory), Analele Academiei, Române, serie III, XX, 4, 1945.
7. *Aerodinamica* (Aerodynamics), ed. tehnică, Buc., 1951
   Tragflügeltheorie (in German), Verlag Technik, Berlin, 1954
   Aerodinamika krila samoleta (in Russian), Izd. Akademii Nauk SSSR, Moskva, 1956.
8. *Mecanica Fluidelor* (Fluid Mechanics), collab. T. Oroveanu, Edit. Academiei RSR, vol. 1, 1952, vol. II, 1955.
9. *Miscări conice in regim supersonic* (Conic movements at supersonic speeds), in Chinese, Acad. de stiinte Beijing, 1955.
10. *High-Speed Aerodynamics* (in English), Buch. Ed. techn., 1956
    Aerodinamica vitezelor mari (in Romanian), Ed. Acad. 1957
    Aerodinamika bolshih skorostei (in Russian), Izd. Akad. Nauk SSSR, Moskva, 1960.
11. *Wing Theory in Supersonic Flow* (collaborators, Dan Mateescu and A. Măstase), Pergamon Press, Oxford, 1969.
12. *Dinamica fluidelor incompresibile* (Incompressible Fluid Dynamics), Bucharest, Edit. Academiei, vol. 1, 1981; vol. 2, 1982.

Ten scientific reports, published in Comptes-Rendus of the Academy of Sciences of Paris.

Eighty-four scientific reports presented at the Romanian Academy or published in the Academy's journals.

Thirty-one scientific articles or papers presented at various international congresses or published in foreign journals.

Thirty-five scientific notes and international articles published in Romania or abroad.

Some important technical or experimental achievements (aircraft, factories, wind-tunnels, aerodynamic laboratories, etc.).

An Ample Prestige in His Country and Abroad

The assignment and award of Romanian and foreign distinctions, the appointment to various important functions, and also, the election to many prestigious international scientific societies and their committees, obviously speaks to Professor Carafoli's merits.

A full professor since 1928, a member of the academy since 1948 and chairman of its technical sciences department, honored scientist, Laureate of the State Prize, awarded the Order of Labor first class, the Star of Republic second class, and Scientific Merit first class, Professor Carafoli is widely known abroad, too. Evidence of this is the international acknowledgement of his lifetime activity in education, research and aerospace development; Professor Carafoli is an Honorary Fellow of the Royal Aeronautical Society of London, a member of the International Academy of Astronautics, a corresponding member of the Academy of Sciences and Arts of Toulouse (France), a former member of the Executive Committee of the International Council of the Scientific Union and of the Council of GAMM (Society for Mathematics and Applied Physics), a member extraordinary of the Hermann Oberth Astronautische Gesellschaft and foreign member of the Braunschweigische wissenschaftliche Gesellschaft (the last three in the Federal Republic of Germany). Professor Carafoli received numerous foreign orders and distinctions: the "Louis Bréguet prize" (1927), the "Médaille d'Monneur-argent" awarded by the French National Society for Progress (1928), the "Paul Tissandier diploma" of the International Aeronautical Federation (1956), the "Silver Medal" awarded by the French National Society for the Promotion of Research and Inventions (1967), the outstanding "Carl Friederich Gauss Medal" (1970), awarded every year by the Braunschweigische Gesellschaft to but one scientist in the world.

Acknowledgements

The author would like to express his gratitude to Academician Professor Elie Carafoli for assistance and indications granted in his career. Also, the support for preparation and presentation of this paper must be accorded to the National Council for Science and Technology (41 Rome strt, Bucharest 1), to the Academy of the Socialist Republic of Romania (125 Calea Victoriei, Bucharest, 1) and to the National Center of the Romanian Aircraft Industry (133 Calea Victoriei, Bucharest 71.102, Romania).

Bibliography

Milne-Thomson, Sir V. *Theoretical Aerodynamics*, London: Macmillan Publishing House, Fourth Edition.

Oroveanu, Teodor. "Cuprilejul aniversării academicianului Elie Carafoli" (At the anniversary of Professor Elie Carafoli) *Studii si cercetări de mec. aplic.* (Applied Mechanics Review), Vol. 40, No. 6, 1981.

Zăgănescu, Florin. *De la Icar la cuceritorii Lunii* (From Icarus to Moon Conquerors), Albatross Publishing House, Bucharest, 1975.

Zăgănescu, Florin. "Ailes roumaines" (Romanian Wings), *Romanian Review*, Vol. 35, Nos. 6-7, 1981.

Zăgănescu, Florin and Gheorghui C., "Historical Data on the Romanian Industry," Aviation, Techn., Bucharest, 1981.

Zăgănescu, Florin. "Elie Carafoli," *Contemporanul Journal*, No. 14 (1847), 2 April 1982.

Zăgănescu, Florin. "Elie Carafoli at 80," *Romania Today*, No. 8 (333), August 1982.

Chapter 18

Ary Szternfeld: Pioneer of Space Navigation[1]

Władysław Geisler[2]

Ary Szternfeld was born in 1905 in the little town of Sieradz in Poland. He attended secondary school in Łódź, a city nearby, and then started his studies at the Jagiellonian University in Cracow. Afterwards he moved to France to continue his studies at the University of Nancy, where he obtained the degree of mechanical engineer in 1927, finishing second among 31 persons who passed the examination. From his early youth he was interested in mechanics, astronautics and space flight, and he was sure that soon people would begin to explore space.

In working on his doctor's degree at the Sorbonne University in Paris, he wrote his thesis on interplanetary flight. While working on this research, which took him three years, he encountered repeated attempts by his academic committee to dissuade him from dealing with astronautics and the movement of heavenly bodies. Committee members considered the subject impractical, that it involved too distant a future, and they offered other dissertation topics that carried prestigious scholarships. In spite of this, Szternfeld preferred to abandon obtaining his doctorate in favor of continuing the work of his life. Thus, he left Paris without a degree and went to stay at his parents' home in Łódź. For some time thereafter, he lived there in complete isolation from the world, working up to 20 hours a day on the subject that most fascinated him.

Using only logarithmic tables, or a simple manual counting machine borrowed for the weekends, he made thousands of laborious calculations and formulated his conclusions, as shown in Table 1. Basing his work on already known theoretical principles of the mechanics of rocket motion and on the calculations carried out by previous theoreticians and pioneers, Szternfeld thoroughly checked those earlier results and somtimes

---

[1] Presented at the Fifteenth History Symposium of the International Academy of Astronautics, Rome, italy, 1981.

[2] Polish Astronautical Society, Katowice, Poland.

even discovered mistakes in the work of his predecessors. Accordingly, he developed and polished up new elements of rocket flight theory by derivation of dozens of his own formulas and gained an indisputable priority in the following fields and subjects:

1. Theory of multi-stage rockets and movement of such rockets in the atmosphere.
2. Method of increasing rocket flight ceiling, and, in aviation, of aircraft flights along the ballistic curve—with motors switched off.
3. Method of determining rocket fuel consumption during vertical flight through the atmosphere.
4. Principle of orbital or space rocket lift-off from the Earth surface vertically, followed by a curvature to a horizontal flight path tangent to the Earth, only after passing through the most dense layers of the atmosphere—generally used today for the reduction of fuel consumption necessary to overcome the force of gravity and air resistance.
5. Creating optimum conditions for maximum height of rocket ascent depending on the initial stock of propellant.
6. Optimum pressure in the combustion chamber.
7. Maximum rocket kinetic energy.
8. Formula for instantaneous rocket efficiency.
9. Total efficiency formula of rocket moving through a resisting medium.
10. Formula for total dynamic rocket efficiency depending on useful mass.
11. Maximum impulse of exhaust gases.
12. Calculation method of stratospheric rocket.
13. Laying out the flight path from the elliptical orbit to another heavenly body—the Sun or a planet—which allows for a reduction by half of the energy used in the case of a flight along the classical semi-ellipse.
14. Designing of an instrument for measuring the distance covered by the rocket, an odograph.
15. Optimum circum-Earth orbits, trajectories leading to the Moon and planets, and rocket speeds essential for putting artificial satellites around the Earth or space probes on interplanetary paths.
16. Calculation of path parameters for interplanetary probes with a foreseen automatic return flight to Earth, after a certain time.
17. Determination of the first space velocity essential for entering an artificial satellite of the Earth into orbit, and calculation of further minimum space velocities necessary for escaping from the Earth's gravitation range, for leaving the solar system, etc.
18. Ratio of achieved velocities and path distances-covered by the rockets with constant acceleration in conditions of Earth gravitation and beyond.
19. Minimization of fuel consumption during the flight from one point of the Earth to another, along a path in the form of a single elliptical arc or consisting of several arcs.
20. Determination of the nearest practical circular orbit for an artificial satellite of the Earth—an orbit located 200 km from the Earth. Also the perigee of elliptical orbits should not be nearer to the Earth than 200 km.
21. Perfection of Hohmann's spiral trajectory for circling Venus.

22. Proving the possibilities of reaching the stars during the lifetime of one generation, based on Einstein's theory of relativity.
23. Examples of paradoxes in rocket engineering: the possibility of achieving more favorable flight results by using, in given conditions, a pressure reduction in the combustion chamber, reduction of used propellant mass, increasing the rocket load, or by introducing intervals in rocket motor operation.
24. Conception of a space suit which would increase body resistance to acceleration, later used in aviation and spacesuits developed in the 1960s.

Table 1

No.	Orbit no. in Sztern-feld's book	Name of flight /of object/	Lift-off date	Parameter determination	In Sztern-feld's book	Executed quantity	Deviation %
1	41	1st Artificial Planetoid USSR "Lunnik 1"	2.1.59	Perihelium/in AU/ Aphelium/in AU/ Big orbit semi-axis in AU Duration of one circle in years	1.000 1.321 1.160 1.250	0.979 1.319 1.149 1.232	2.14 0.15 0.96 1.46
2	36	1st Artificial Planetoid USA "Pioneer 4"	3.3.59	Perihelium/in AU/ Aphelium/in AU/ Big orbit semi-axis/in AU/ Duration of one circle in years	1.000 1.146 1.073 1.111	0.982 1.136 1.059 1.090	1.83 0.88 1.32 1.93
3	31	2nd Artificial Planetoid USA "Pioneer 5"	11.3.60	Perihelium/in AU/ Aphelium/in AU/ Big orbit semi-axis/in AU/ Duration of one circle in years	0.805 1.000 0.902 0.857	0.803 0.993 0.898 0.851	0.25 0.70 0.44 0.70
4	28	Interplanetary probe USSR Wenus 1 "ASM-2"	12.2.61	Perihelium/in AU/ Aphelium/in AU/ Big orbit semi-axis/in AU/ Duration of one circle in years	0.724 1.000 0.862 0.800	0.709 1.010 0.860 0.797	2.12 1.00 0.23 0.38
5	37	3rd Artificial Planetoid USA "Ranger 3"	26.1.62	Perihelium/in AU/ Aphelium/in AU/ Big orbit semi-axis/in AU/ Duration of one circle in years	1.000 1.163 1.082 1.125	0.995 1.164 1.074 1.113	0.50 0.09 0.74 1.08
6	46	3rd Artificial Planetoid USSR "Mars 1"	1.2.62	Perihelium/in AU/ Aphelium/in AU/ Big orbit semi-axis/in AU/ Duration of one circle in years	1.000 1.621 1.310 1.500	1.000 1.600 1.300 1.482	0.00 1.31 0.77 1.21

Some of these problems were developed and studied more thoroughly by Szternfeld in his later publications. However, his basic ideas and calculations appeared in the work that he prepared in Paris and completed early in his career in Łódź, called *Initiation à la Cosmonautique*, which was initially published in French in 1933. In his first book, which became his fundamental work, Szternfeld also dealt with many other different problems such as the history of rocket engineering development, the quantities of food and air essential for crews on long-lasting space expeditions, the idea of constructing space mirrors and orbiting settlements, interplanetary signaling, the problem of visibility of artificial satellites from Earth, and the like.

At first Szternfeld found little understanding for his ideas. After completing *Initiation à la Cosmonautique*, on 6 December 1933, Szternfeld presented its principal points at a meeting of the Polish Society of Astronomy at the Astronomical Observatory in Warsaw. In general, his ideas were received rather skeptically even though no errors in argumentation or calculations were raised.

This began to change later that same month, however, when Szternfeld went to Paris and presented his work to the French Astronomical Society where it met with greater interest and approbation. Two chapters of his treatise, one on the measurement of distance covered by a rocket by means of an odograph, and the other on the idea of applying a paradoxical but more economical trajectory—leading from Kepler's orbit to the central body—were presented during sessions of the French Academy of Sciences in Paris. This was an unprecedented event due to the completely new subject matter, never before discussed at meetings of this Academy. After discussion, the author's theses were accepted as correct. Szternfeld's lecture, entitled "Some New Ideas in Astronautics," which was delivered on 2 May 1934 in Descartes' Great Assembly Hall at the Sorbonne University in Paris, proved to be a great success for the young engineer from Poland.

Robert Esnault-Pelterie, the "father" of French astronautics, sincerely congratulated him, and Hermann Oberth and Walter Hohmann paid their tribute in writing, as did Konstantin E. Tsiolkovsky from the Soviet Union. A month later Ary Szternfeld received, also in France, the international astronautical R.E.P. Hirsch prize. However, he could not find a Polish publisher for his first book, even though he tried diligently.

In 1935 he accepted an invitation to visit Moscow, where leaders in the Soviet scientific community expressed an interest in his works. He went there with his wife, hoping to find the right conditions for continuing his work in astronautics and to meet Tsiolkovsky, with whom he had been corresponding. The unexpected death of Tsiolkovsky in September 1935, however, made any personal contact impossible. Nevertheless, Ary Szternfeld had the opportunity to remain and work in RNII (Scientific and Research Institute of Jet Propulsion) in Moscow, together with designers and constructors of Soviet rockets including S. P. Korolev, W. P. Glushko, M. K. Tichonorawow, G. E. Łangiemak and J. Pobiedonoscew. This gave him an opportunity to get acquainted with rocket construction technology and its application in practice, thus enhancing his theoretical knowledge with practical experience. This experience also allowed him to supplement the contents of his book with a chapter on rocket engineering, in particular on multistage rockets and their flight into the atmosphere. G. E. Łangiemak, Szternfeld's direct superior and friend, also translated his book from French into Russian and helped considerably in securing its publication in the Soviet Union in 1937.

In 1936 Ary Szternfeld and his wife became Soviet citizens. However, the following year he had to leave RNII to begin working in Moscow's machine industry, where he performed various other design and construction labor as mechanical engineer. During the war, 1941-1944, he also taught in a metallurgic technical school in Sicrow in the Ural Mountains. He could never during this period occupy himself professionally with rocket engineering and astronautics. As it was his passion, Szternfeld constantly tried to find a way to devote full time to rocketry, and he kept sending applications and suggestions to different institutes, stressing the need for taking up research work on the development of rocket engineering. He urged preparations be made for the forthcoming exploration of space, which, according to him, would become reality in the near future. All his endeavors proved to be without success. He could only continue his works and calculations in astronautics by staying up late at night and doing them in his leisure time. He elaborated many new problems from the sphere of intercontinental and space rockets launched at different angles in relation to the Earth. He also calculated many paradoxical, but economic, trajectories leading to other planets and heavenly bodies in the solar system.

These calculations won him the name, "Lord Paradox of Astronautics," similar to Oscar Wilde who was called "Lord Paradox of Literature." To show these paradoxical, in appearance, solutions to different problems which arise in rocket engineering and astronautics, the following examples can be given:

1. In certain cases the altitude of rockets flying above the Earth, after lifting-off horizontally, can be greater than those lifting-off vertically, with the same amount of fuel consumed.
2. Higher altitude of flight can be obtained by a rocket starting from Earth by temporarily reducing the pressure in the combustion chamber, or even by interrupting, for short periods, the process of fuel combustion.
3. Additional loading of the rocket during lift-off may give more favorable parameters during flights in the atmosphere at low *altitudes* than a rocket that is less loaded.
4. The increase of flight velocity of a rocket traveling along an elliptical orbit can cause it to reach its destination later than the rocket which is flying at a lower speed to the same destination point. A determined flight velocity increase gives a bigger increment of flight path length—the ellipse becomes proportionally longer to the square of velocity increase.
5. There also occur opposite cases. As an example, Szternfeld gave a precisely calculated flight from the Earth to the Moon along a semi-elliptical path, at a speed similar to that used today along that route. And so, in the case of minimum rocket flight acceleration—by only 0.8 percent—it will be possible to reduce the duration of the trip by half, or by 50 percent. The reduction of speed by that *same* small fraction of a percent (0.8 percent) will cause the rocket not to reach the Moon but to start falling towards the Earth.
6. Let us assume that a rocket lifts off at 12:00 noon from a given point of the Earth in a eastward direction and achieves escape velocity. Due to this velocity the rocket will start moving away from the Earth, so as never to return to it. If 12 hours later, at mid-night, an identical rocket starts from that same point in an eastward direction,

however with greater speed, it would seem that it would soon catch up with the first. Nevertheless, this will never happen. The first rocket will be moving away from the Sun aiming for the orbits of Mars and Jupiter, whereas, the second shall be heading for the Sun, after crossing the orbits of Venus and Mercury. This is so, since in one case there is a summation of the rocket motion velocity as *against the* Earth, and of the Earth around the Sun, whereas in the other case there is a subtraction of these two velocities. Thus their are completely different shapes of the trajectories of each rocket.

7. The shortest route to many destinations in space does not necessarily have to be the cheapest. It is usually connected with larger fuel consumption than is a route which is longer, but which has been chosen in such a way that most of it can be covered due to the action of the gravity force and with rocket motors switched-off. The motors *are only* switched on for a short distance, thus allowing for a considerable saving of fuel.

Szternfeld proved to be a master in discovering many such unexpected, and at the same time favorable, solutions, because of his unique understanding of the laws of physics and mechanics and their linking together through his remarkable insight and superb command of mathematics.

Szternfeld included several of these paradoxes in his first Russian book, *Wwiedienije w Kosmonawtiku* (Introduction to Astronautics), which was in general very well received in the U.S.S.R. M. K. Tichonoravov, the builder of the first Soviet rocket run by liquid propellants in 1938, among others, wrote the following words:

> All calculations made by Szternfeld are much more accurate than those of Oberth, Esnault-Pelterie and Hohmann. Whereas, one very important thing should be stressed at this point, and i.e., that A. Szternfeld's calculations from the point of view of interplanetary travel are much better than those of all the other scientists . . . . Whilst all scientific investigators of interplanetary flight routes did not go beyond the solar system, A. Szternfeld, by applying the relativity theory (which Esnault-Pelterie tried to do before, however, without success) calculated the routes to the nearest stars and obtained quite good results . . . . For the first time the theory of the multi-stage rocket was developed by A. Szternfeld in his book . . . . The author, also for the first time, elaborated the theory of interplanetary signaling.

The success that the 1938 book enjoyed, especially in the Soviet Union, did not make Szternfeld's life much easier. Not being able to work professionally in the field for which he had such a passion, he had to take up jobs in different technical and industrial establishments in order to earn a living and maintain his family. Even so, he continued to publish the results of his laborious calculations and studies in astronautics and rocket engineering, virtually all conducted during his free time, in articles and in popular scientific books. He tried to present the contents of these books, often containing the results of complicated scientific research, in the most comprehensible manner possible—and thus, he soon became well-known, first in the U.S.S.R. and then abroad.

In 1949 his book, *Poliot w Mirowoje Prostranstwo* (Flight into Space), was published in 50,000 copies. It quickly sold out. In 1950, Szternfeld's booklet, *K. E. Ciołk-*

owskij i Mieżpłanetnyje Putieszestwija (K. E. Tsiolkovsky and Interplanetary Travel), was published. In 1955 his second popular scientific book, entitled simply *Mieżpłanetnyje Polioty* (Interplanetary Travel), was issued in a large press run of 100,000 copies and had to be reprinted in an equal number of copies. His next book, *Iskusstwiennyje Sputniki Ziemli* (Artificial Satellites of the Earth)," was published in 1956. After *Sputnik I*, it was supplemented and again published in 1958 under the title *Iskusstwiennyje Sputniki* (Artificial Satellites) in a total of 150,000 copies and became in the Soviet Union something of an encyclopedia of knowledge on artificial satellites. These books presented a running compilation of Szternfeld's thinking on the calculations of orbits; principles of motion; types of artificial satellites (among them geostationary satellites); essential orbital velocities; visibility of artificial satellites from the Earth; lift-off and landing of these craft; carrier rockets; life of the crew on board artificial satellites of the Earth; future satellites; and space stations and laboratories on Earth's orbit. The 1958 book by Szternfeld, as well as his first work, *Wwiedienije w Kosmonawtiku* (Introduction to Astronautics), became obligatory reading for cosmonaut candidates in the U.S.S.R.

Iskusstwiennyje Sputniki (Artificial Satellites) was also published in other countries of Europe and in the United States, selling more than 200,000 copies, and gaining for Szternfeld an audience for his ideas outside the Soviet Union. Willy Ley, in his introduction to this book, when published in the United States under the title *Soviet Space Science*, wrote: "Considering its qualities, the book is one of the best reviews on the subject ever published in any country. . . ."

Szternfeld's next book *Ot Iskusstwiennych Sputnikow k Mieżpłanietnym Poliotom* (From Artificial Satellites to Interplanetary Flights) had a more popular character and was also widely read. It was first published in 1957 in Moscow, in 100,000 copies, and a second edition appeared in 1959. Szternfeld followed this in 1960 with the publication in Bratislava, Czechoslovakia, of a collection of short popularized scientific stories entitled *Človek Pokoruje Vesmir* (Man Conquers Space) and in 1967 with a book called *Sladami Kosmonautycznych Koncepcji z lat 1929-1936* (Following the Astronautical Conceptions through the Years, 1929-1936), published in Szczecin, Poland, in the Polish language.

Szternfeld's publishing efforts over the years were impressive. Books written by Szternfeld have had a total of 85 printings in 36 languages, and total sales of more than one million copies in 29 countries of Europe, America, Asia, Africa, and the nine Soviet Republics. Apart from his books, Ary Szternfeld published more than 400 scientific articles and historical essays in many professional and scientific journals and in popular magazines of many nations, each adding to either the body of scientific knowledge or enhancing the popular perception of it. He also wrote for a variety of daily newspapers, mainly in the U.S.S.R., Poland, and France, but also in Italy, Czechoslovakia, East Germany, Bulgaria, Romania, the United States, Great Britain, Belgium, West Germany, Sweden, Switzerland, Finland, Portugal, Mexico, Brazil, Argentina, Korea and China. His most widely circulated article, "On Interplanetary Travel," was picked up by the United Press Agency and run in more than 2,500 dailies in Western Europe and both Americas.

More than a popularizer, although that was a valuable aspect of his work, Szternfeld remained professionally active throughout his career and participated in numerous conferences of scientific institutes in the Soviet Union, France, and Poland. For example, on 25 August 1945 he gave a lecture on the consumption of rocket propellant during a flight through the atmosphere at a predetermined angle to the Academy of Sciences of the U.S.S.R. in Moscow. He also gave the pathbreaking scientific papers:

1. 3 March 1951—during the 3rd Meteorite Conference in Moscow on the Tunguska Meteorite and on the dead navigation seasons in interplanetary space.
2. 26 May 1955—in the U.S.S.R. Academy of Sciences in Moscow "On the possibility of jump-reduction of space flight duration with a uniform increase of rocket velocity."
3. 4 September 1956—during a meeting of the Polish Astronautical Society at the Technical University in Warsaw, a paper on "Some Problems Connected with Sending Artificial Satellites of the Earth and with Choosing Interplanetary Trajectories."
4. 13 April 1961—in the Lorrainian Scientific Society in Nancy, France, a paper on "Ideas of Priority in Astronautics." This lecture was also delivered for a second time, in a more complete form, at the University in Warsaw on 27 July 1964.
5. 29 November 1976—lecture at the Technical University of Warsaw in Poland, entitled "Paradoxes in Astronautics".

Ary Szternfeld was also the author of several proposals for patents submitted in Belgium and the Soviet Union, referring among others to a remotely controlled robot that could perform difficult and dangerous-to-manage tasks both on Earth and in space flight. From 1953 to 1960 he was editor of the department of "Astronautics" in the bulletin, *Refieratywnyj Zurnał* (Journal of Abstracts), published by the Academy of Sciences of the Soviet Union. At the beginning of 1954, together with General Krasnojurczenko and Colonels Warwarow and Chlebcewicz, as well as several others, he founded the Astronautic Section attached to the Czkałow Aeroclub in Moscow, and for several years he conducted periodic seminars there on space navigation.

Even with these accomplishments Ary Szternfeld believed he had to fight for astronautics to be accepted by the scientific world as a legitimate field of human activity. He acted stubbornly and persistently for more than thirty years as an advocate of astronautical research, seeking to overcome the technical difficulties and human indifference. He did live, however, to see what he viewed as a great vindication—human space flight—and even more, the Yuri Gagarin flight in 1961 used many of the mathematical models Szternfeld had calculated and proved a quarter of a century earlier.

In 1974 his fundamental work *Wwiedienije w Kosmonawtiku* (Introduction to Astronautics) was published for the second time in the U.S.S.R., without any substantial changes from its first edition in 1937. It did, however, contain numerous additional data on space flight and on experiments carried out in the meantime, all of which confirmed the author's original theoretical expectations of 40 years earlier.

In the foreword to that second edition, the Academician Professor Dr. Jurij Pobiedonoscew wrote: "Not every scientist has the honor to have his book published again more than 30 years after its first edition. The second edition is an act of gratitude by a new generation of scientists to the pioneer of astronautics, who elaborated its principles at the time when many of the people going into space today were not yet born".

This Soviet scientist, with time, lived to receive international fame and several high distinctions and scientific honors of the Soviet Union, France, and Poland. Sztern-feld obtained three Honoris Causa degrees—in 1961 in physical and mathematical sciences from the University of Nancy, in 1965 in Technical Sciences from the Academy of Sciences of the U.S.S.R. in Moscow, and in 1977 at the National Polytechnical Institute of Lorraine in France. The Astronomical Council of the Academy of Sciences of the U.S.S.R. gave him an honorary diploma in 1965 for his many years of studies and popularizing scientific knowledge in the field of astronautic works, as well as a commemorative award of "Meritorious Observer of Artificial Satellites." That same year, Szternfeld also obtained the honorary title of "Meritorious Worker Actively Engaged in Science and Technique of the RFSRR" for merits in the field of astronautics.

In addition to awards in the Soviet Union, in 1960 the Lorrainian Scientific Society in Nancy, France, awarded to the Soviet pioneer the rank of honorary member, and in 1966 similar status was also awarded by the Academy of Sciences and the Scientific Society in Nancy. The Polish Astronautical Society gave this scientist of Polish origin an honorary award, and his home town, Sieradz, commemorated him on 17 May 1963, by conferring on him the distinction of "Honorary Citizen of Sieradz." On 25 January 1965, the Municipal Council of Sieradz also conferred on him the title of "Merited for the Town of Sieradz." Also in January, he received the Robert Esnault-Pelterie Medal from France, and in 1978 he was awarded the Gold Medal of the Eastern Industrial Society in France. Two years later, on 5 July 1980, Ary Szternfeld died in Moscow at the age of 75.

Index

Index to all American Astronautical Society Papers and Articles

1954 - 1992

This index is a numerical/chronological index (which also serves as a citation index) and an author index. (A subject index volume will be forthcoming.)

It covers all articles that appear in the following:
Advances in the Astronautical Sciences (1957 - 1992)
Science and Technology Series (1964 - 1992)
AAS History Series (1977 - 1992)
AAS Microfiche Series (1968 - 1992)
Journal of the Astronautical Sciences (1954 - September 1992)
Astronautical Sciences Review (1959 - 1962)

If you are in aerospace you will want this excellent reference tool which covers the first 35 years of the Space Age.

Numerical/Chronological/Author Index in three volumes,
Ordered as a set:
Library Binding (all three volumes) $120.00;
Soft Cover (all three volumes) $90.00.

Ordered by individual volume:
Volume I (1954 - 1978) Library Binding $40.00; Soft Cover $30.00;
Volume II (1979 - 1985/86) Library Binding $60.00; Soft Cover $45.00;
Volume III (1986 - 1992) Library Binding $70.00; Soft Cover $50.00.

Order from Univelt, Inc., P.O. Box 28130, San Diego, California 92198.

Numerical Index[1]

Volume 11, AAS History Series, *History of Rocketry and Astronautics*, (IAA History Symposia, Volume 6)

(Proceedings of the Fifteenth and Sixteenth History Symposia of the International Academy of Astronautics, Rome, Italy, 1981; Paris, France, 1982)

AAS 91-251 On the Design of N. I. Kibal'chich's Flying Machine (on the 100th Year of its Development) (IAA 81-317), A. M. Vinitskii

AAS 91-252 The Origins of Inertial Navigation in Space (IAA 82-274), A. Yu. Ishlinsky

AAS 91-253 Evolution of Space Fiction in Film (IAA 82-279), Frederick I. Ordway, III

AAS 91-254 Rocketry Personnel Training in the U.S.S.R. (1924-1936) (IAA 82-278), E. K. Moshkin and A. F. Nistratov

AAS 91-255 The Supersonic Wind Tunnel Installations at Peenemünde and Kochel and Their Contributions to the Aerodynamics of Rocket-Powered Vehicles (IAA 81-321), Rudolf Hermann

AAS 91-256 Origins of Magnetospheric Physics (IAA 81-320), James A. Van Allen

AAS 91-257 Reaction Motors, Inc.: A Corporate History, 1941-1958, Part I, Institutional Developments (IAA 82-277), Frederick I. Ordway, III and Frank H. Winter

AAS 91-258 Reaction Motors, Inc.: A Corporate History, 1941-1958, Part II, Research and Development Efforts (IAA-82-277), Frank H. Winter and Frederick I. Ordway, III

AAS 91-259 Early Experiments with Erosive Burning in Solid Rockets (IAA 81-322), J. M. Murphy and J. R. Osborn

AAS 91-260 Fundamental Scientific Questions in the Early Period of Rocket Propulsion Development (IAA 82-275), Martin Summerfield

---

[1] IAA papers presented in 1981 and 1982. AAS numbers have been assigned for identification purposes.

AAS 91-261 Scientific Foundations for Implementation of Manned Space Flight (IAA 81-325), V. I. Yazdovsky and Y. Vedeshyne

AAS 91-262 The History of Extravehicular Activity (EVA) in U.S. Human Space Flight (IAA 81-322), R. Scott Millican

AAS 91-263 Space Research in Poland After 1958 (IAA 81-318), M. Subotowicz

AAS 91-264 Konstantin Eduardovich Tsiolkovsky and the Present Times (IAA 82-276), A. A. Kosmodemjansky

AAS 91-265 Nikolai Alexeyevich Rynin (1877-1942), Soviet Astronautical Pioneer: An American Appreciation (IAA 81-319), Frank H. Winter

AAS 91-266 The Contribution of Robert Esnault-Pelterie to Astronautics (IAA 82-283), P. Contensou

AAS 91-267 Romanian Professor Elie Carafoli: Fifty-Five Years of Devotion to Modern Aeronautics and Astronutics (IAA 82-280), Florin Zăgănescu

AAS 91-268 Ary Szternfeld: Pioneer of Space Navigation (IAA 81-302), Władysław Geisler

AAS 91-269 to -280
Not Assigned

Author Index[1]

[1] For each author the paper number is given. AAS numbers for these IAA history papers presented in 1981-1982 have been assigned for identification purposes. The page numbers refer to Volume 11, *AAS History Series*.

Publications of the American Astronautical Society

Following are the principal publications of the American Astronautical Society:

Journal of the Astronautical Sciences (1954 -): Published quarterly and distributed by AAS Business Office, 6352 Rolling Mill Place, Suite #102, Springfield, Virginia 22152. Back issues available from Univelt, Inc., P.O. Box 28130, San Diego, California 92198.

Space Times (1986 -): Published bi-monthly and distributed by AAS Business Office, 6352 Rolling Mill Place, Suite #102, Springfield, Virginia 22152.

AAS Newsletter (1962 - 1985): Incorporated in *Space Times*. Back issues available from AAS Business Office, 6352 Rolling Mill Place, Suite #102, Springfield, Virginia 22152.

Astronautical Sciences Review (1959 - 1962): Incorporated in *Space Times*. Back issues still available from Univelt, Inc., P.O. Box 28130, San Diego, California 92198.

Advances in the Astronautical Sciences (1957 -): Proceedings of major AAS technical meetings. Published and distributed for the American Astronautical Society by Univelt, Inc., P.O. Box 28130, San Diego, California 92198.

Science and Technology Series (1964 -): Supplement to *Advances in the Astronautical Sciences*. Proceedings and monographs, most of them based on AAS technical meetings. Published and distributed for the American Astronautical Society by Univelt, Inc., P.O. Box 28130, San Diego, California 92198.

AAS History Series (1977 -): Supplement to *Advances in the Astronautical Sciences*. Selected works in the field of aerospace history under the editorship of R. Cargill Hall. Published and distributed for the American Astronautical Society by Univelt, Inc., P.O. Box 28130, San Diego, California 92198.

AAS Microfiche Series (1968 -): Supplement to *Advances in the Astronautical Sciences*. Consists principally of technical papers not included in the hard-copy volume. Published and distributed for the American Astronautical Society by Univelt, Inc., P.O. Box 28130, San Diego, California 92198.

Subscriptions to the *Journal of the Astronautical Sciences* and the *Space Times* should be ordered from the AAS Business Office. Back issues of the *Journal* and all books and microfiche should be ordered from Univelt, Incorporated.